Ecological Transition in Education and Research

Ecological Transition in Education and Research

Edited by

Hassan Ait Haddou
Dimitri Toubanos
Philippe Villien

iSTE

WILEY

First published 2021 in Great Britain and the United States by ISTE Ltd and John Wiley & Sons, Inc.

ISTE Ltd
27-37 St George's Road
London SW19 4EU
UK

www.iste.co.uk

John Wiley & Sons, Inc.
111 River Street
Hoboken, NJ 07030
USA

www.wiley.com

Library of Congress Control Number: 2021945501

British Library Cataloguing-in-Publication Data
A CIP record for this book is available from the British Library
ISBN 978-1-78630-716-3

MIX
Paper from responsible sources
FSC
www.fsc.org
FSC® C013604

Contents

Introduction

This book is dedicated to the theme of climate change, an urgent issue in Ecological Transition, particularly for architecture and landscape schools.

The scientific realities and consequences of this phenomenon are becoming increasingly well-known. It has been established that the mitigation and adaptation of emissions (particularly through the reduction of GHG emissions) directly impacts architecture, urban planning and landscape areas. Climate change must surely be put into relation with other environmental and social crises, depending on the multiple geographical and social situations of the territories involved.

The research work presented in this multidisciplinary collective work under the direction of Hassan Ait Haddou, Dimitri Toubanos and Philippe Villien aims to contribute to the recognition of the consequences of climate change and raise consciousness about mitigation and adaptation actions. In opposition to "climate skepticism", the authors find inspiration in action-research, educational innovations and concrete transition practices, embracing different schools of thought. The challenge is to promote educational practices and research on climate change, to provide food for thought and action relying on the diversity of architectural and landscape responses to global warming and extreme episodes.

These contributions are the result of an international meeting organized by the scientific and educational network "EnsaÉco" and the LIFAM laboratory at ENSA Montpellier, not only combining the knowledge and experiences of "transitioners", but also making a significant contribution to the development of adaptive actions, thanks to architecture and landscape.

Awareness about the challenges related to the ecological transition at higher education institutions continues to grow in France and at an international level. Indeed, rising to these challenges will not be possible without a real commitment from the teaching and research spheres. It is against this background that a scientific and educational network committed to the teaching of ecological transition in architecture and landscape schools, called "EnsaÉco", was founded in November 2016.

Truly committed to this task, the scientific and educational network EnsaÉco initiated a series of scientific events around teaching and research on the themes of sustainable development. A series of meetings followed, bringing together the contributions from teachers, researchers and students from all of the ENSA(P) institutions in France, as well as European and internationally renowned scientific personalities.

Thus, in 2017, in Lyon, the first meetings of the network made it possible to federate the members of the community of teacher-researchers and to raise awareness of plural activities for teaching in architecture and landscape schools[1]. They resulted in the launch of an "Appeal for the teaching of transition in architecture and landscape schools", or "Lyon Appeal"[2].

The second meeting of the EnsaÉco network, which took place in 2018 in Nancy[3], made it possible to debate a series of measures, called "tilting measures", for the teaching and research of ecological transition in architecture and landscape schools.

In 2019, the EnsaÉco network desired to continue this reflection and open up a key question for architecture and landscape schools: that of scientific research, beyond the scope of teaching. It is in this spirit that the third meetings of the network took place in Montpellier, with a strong involvement from the *Laboratoire Innovation Formes Architectures Milieux* (LIFAM, Laboratory of Innovative Forms in Architectural Environments), the historical laboratory from ENSA Montpellier.

The major contributions of these meetings were dedicated to the articulation between teaching, research and professional practice on ecological transition. The particular interest in research on the theme of ecological transition follows various initiatives from the Ministry of Culture to support research related to this topic, starting with the research program "*IgnisMutatRes*: thinking architecture, the city and landscapes through the prism of energy" (2011–2015)[4], followed by the

1 For more information, see: http://ensaeco.archi.fr/manifestations/actes-lyon-2017/.

2 For more information, see: http://ensaeco.archi.fr/appel-de-lyon/.

3 For more information, see: http://ensaeco.archi.fr/manifestations/2018-nancy-_mesures-basculantes/.

4 For more information on the "*IgnisMutatRes*" program, see: http://www.culture.gouv.fr/Thematiques/Architecture/Formations-Recherche-Metiers/La-recherche-architecturale-urbaine-et-

program "20th century architecture, project material for the sustainable city of the 21st century" (2016–2020)[5]. At the same time, a significant amount of specific funding was allocated to research projects on ecological transition, either through the Ministry of Culture, PUCA (*Plan Urbanisme Construction Architecture*), ADEME (*Agence de Maitrise de l'Energie*, now called *Agence de la Transition Ecologique*), ANR (*Agence Nationale de la Recherche*) or CDC (*Caisse des Dépôts et Consignation*). All of these actions enabled a proliferation of research subjects feeding the reflection about the post-carbon city, something that the third meetings of the EnsaÉco network particularly wished to promote. This also made it possible to fuel teachings in architecture and landscape schools, while forging links between the academic and professional world.

As a result of these meetings, the EnsaÉco network wished to publish its first scientific work. The challenge of the publication was to promote diverse research projects, enhance knowledge-crossing across researchers, contribute to the development of innovative research and re-establish links between researchers, practitioners and citizens.

Using architectural, urban and landscape research material, EnsaÉco's scientific committee identified several challenges of ecological transition in response to climate change. This question has been tackled following three general axes, voluntarily open to all disciplines, covering a variety of issues:

– *The urban dimension*: this axis focuses on bioclimatic management in an urban context: the dense city, the urban heat island (UHI), the plants in the city, transportation policies, or rehabilitation standards.

– *The territorial dimension*: this axis proposes to shift the focus from the specifically urban question to examine the relationships between the city and the problem of peri-urbanization, particularly exploring the way in which vernacular cultures in rural areas have an heuristic power for the development of architectural models likely to better respond to the challenges of ecological transition. The agricultural question is also key to this axis, as well as the relationship between the territory and the sea.

– *The social dimension*: embodying the third pillar of sustainable development, the social dimension is at the heart of paradoxes, contradictions and (geo)political and socio-economic conflicts, which must necessarily be called into question.

---

paysagere/L-organisation-de-la-recherche/La-politique-incitative-a-la-recherche/Ignis-Mutat-Res-IMR [Accessed October 7, 2018].

5 For more information on "20th Century Architecture", see: http://www.culture.gouv.fr/Thematiques/Architecture/Architecture-et-cadre-de-vie/Architecture-du-XXe-siecle [Accessed October 7, 2018].

The texts collected following our call for contributions deliver initial intentions and insights to clarify the challenges of ecological transition in teaching, research and operational practices. The diversity of the contributions received testifies to an array of interdisciplinary approaches, putting the question of ecological transition at the center of the discussion within a global framework: climate change. The three general axes (urban, territorial and social dimensions) are approached in a transversal manner in the various articles reunited in this corpus.

This book is structured into three main sections, hosting a variety of themes which illustrate the issues raised by the three general axes previously mentioned.

The first section presents the educational processes at the service of ecological transition, by paying particular attention to good practices in terms of the teaching and research of the ecological challenges to be taken up by architects, town planners and landscapers. It is mainly based on teaching experiences across the ENSA (*École Nationale Supérieure d'Architecture*) institutions in France.

The first chapter begins with the conference of one of the pioneers of what is called bioclimatism. André De Herde delivered the general introduction to the three days of EnsaÉco meetings. The author structures his remarks into three main blocks. The first one presents the "bioclimatic", "sustainable" and "smart" concepts proposed by architecture schools to remedy the energy crisis of the 1970s. The second one recalls the reasons for the creation of the "Architecture and Climate" research team in 1980. He concludes his contribution with a state-of-the-art of the teachings of architectural engineers at the University of Louvain-La-Neuve.

De Herde's conference is followed by the testimonies of three teacher-researchers, Anne Coste, Frédéric Dellinger and Théodore Guuinic, who illustrate the engagement of the field of history as an independent discipline and as a privileged dimension of other educational fields for acknowledging the constraints related to climate change. The authors of this work discuss the way in which climate change considerations have shaped the teaching of the history of architecture, town planning and landscape. They also show how the historical dimension makes it possible to put the climatic questions addressed by other disciplines taught at the ENSA(P) into perspective.

This chapter is followed by the contributions of Alain Guez and Antoine Steck from ENSA Nancy, who use intensive teaching to make students aware about the challenges of ecological transition in the Anthropocene era, and invite them to apprehend possible levers of action by working from concrete situations at different territorial and temporal scales.

Also in the context of the feedback from educational processes at the service of ecological transition, the contributions by Emmanuel Doutriaux and Edith Akiki are based on an experience at the interface between teaching and research. The authors work on the environmental question by bringing together their respective expertise, with a special focus on climate change.

The first section of the book closes with the work of Christophe Laurens, Valentin Sanitas, Clément Gaillard, Tibo Labat and Jérémie Buttin, who aim to present the experience of a group of students carried out at the ZAD NDDL (Zone to Defend, Notre-Dame-des-Landes), a territory which articulates many concrete dimensions. The authors suggest a few avenues in order to understand how to represent the contemporary political and ecological issues specific to this struggling territory, by means of representation and architectural design.

The second section of the book discusses ecological transition challenges that research needs to address.

Based on a case study located in a tropical climate (Reunion Island), as part of her final thesis edited by Hassan Ait Haddou, Magalie Técher explores the notion of comfort at elementary schools in humid tropical climates. The authors address questions which are at the core of architectural thought, in a context of ecological and economic crises, with the aim of providing a healthy and sustainable environment for children. They seek to provide food for thought and avenues for improvement in the design of school buildings. The analysis of data resulting from a campaign of indoor atmosphere measurements and a perception survey made it possible to determine optimal comfort zones for students at the different climatic zones of Reunion Island.

The study conducted by Séverine Steenhuyse in Marseille, focuses on the phenomenon of the urban heat island (UHI). The author of this project calls into question the virtuous planning of this city, with the aim of preserving its qualities and contributing to the mitigation of this phenomenon (UHI). The purpose of her work is to find out to what extent urban vegetation can actually be used, first analyzing how the phenomenon of urban overheating works, and discussing its known mitigating factors. Later, the author explores the issues of the urban green and blue network in order to imagine how these two issues could combine their approaches.

In Toulouse, Isabelle Fortuné carried out a study on summer comfort in the face of climate change. The author invites us to take into account the overall cost and the sensations within our habitat if we wish to be in line with the challenges of ecological transition.

A collaborative and multidisciplinary perspective concludes this second section with the contribution of Cécile Fries Paiola, who presents the work of a team of teacher-researchers from ENSA Nancy. Their approach combines lessons from project workshops, scientific mediation actions, and ultimately, joint research projects, in order to find answers to the new constraints of digital, ecological and societal transitions.

The third and last section demonstrates how professionals face the new challenges brought about by major changes. This is precisely what Damien Antoni describes in his chapter. The author proposes to respond to two distinct themes, namely the ways of mitigating global warming and giving testimony of the ecological architectural projects in progress.

With the purpose of dealing with global warming, François Nowakowski describes an educational experience carried out in partnership with the community of municipalities in the Ardèche valley, at *Vallée de la Ligne*. Based on the theoretical framework of the Italian territorialist school, this work has made it possible to identify the dynamics of change at work in the territory, either in the direction of adaptation/mitigation of climate change, or on the contrary, its amplification.

Back in Toulouse, the work of Clément Gaillard focuses on the bioclimatic houses of Jean-Pierre Cordier, who is considered as an important and influential figure of this design process. The chapter aims to understand not only the motivations, the architectural and constructive choices, but also the errors which appear in various projects by Jean-Pierre Cordier built between 1981 and 1986.

The last contribution of this section echoes the !MPACT competition, a contest open to all students of architecture and other training schools related to building, engineering, design and art, both in France and at an international scale.

Through these varied contributions, we can notice the diversity of approaches dealing with the issue of climate change in architecture and landscape schools. This testifies to the interdisciplinarity within schools, thanks to the teachers, researchers, students and administrative staff who work at them. By combining these varied texts, we hope to have contributed to the challenge of integrating ecological transition into the teaching and research departments of architecture and landscape schools.

PART 1

Educational Processes at the Service of Ecological Transition

1

Between Architecture and Climate: A Bioclimatic Approach

1.1. Introduction

When André De Herde takes his place at the microphone, we feel that the air thickens with respect and expectation. "André is one of the pioneers of what is called bioclimatism," Robert Célaire introduces him, also describing him as "professor emeritus, engineer architect, doctor and author of several books on bioclimatic architecture". These words bring me back to the time when I arrived at an agency as a young graduate engineer-architect from Leuven (not to be confused with Louvain-La-Neuve, which is also in Belgium), and my boss presented me with the *Bioclimatic Architecture Guide: Volume 2* as required reading. Without knowing it, André De Herde had a great influence on my architectural approach and production.

He speaks up. His calm voice reveals a great experience in the field. He says he feels honored to be here, in front of the Minister's Representative and the students, and we, in turn, feel honored to be with him.

He unveils the structure of the conference in three parts: the concepts of bioclimatism, sustainable development and smartness; the "Architecture and Climate" research team created in 1980; and the teaching of architectural engineers at the University of Louvain-La-Neuve. He specifies that there will be around 35 numbered slides, for those who may find the conference long. The framework is set, and the conference can start.

This chapter comprises the words of André DE HERDE, transcribed by Marjan SANSEN.

1.2. Bioclimatism

Following the energy crisis, in 1973–1976, architecture schools proposed adding 10 cm of insulation to walls and solar panels on roofs. At no time has the energetic aspect been taken into consideration in the architectural act or the building design.

We offer an alternative to this, one possible answer among other possible answers, without claiming it to be *the* answer or *the* ultimate truth.

Our starting point is the observation that every building that we build or renovate is located somewhere, set within a certain climate. All these climates present positive and negative aspects. It is only humid tropical lowland climates that have exclusively negative aspects. The positive aspects are sometimes only at night, or simply because we are at a certain altitude and this enables us to act to guarantee the comfort of the occupants.

For us, from the design phase to the realization phase, the architectural act has two goals. The first is the search for the best match between the habitat – the building, the climate in which the building is located – and the occupant. The occupant is particularly important, because it is them who bring forward the comfort requirements. The four comfort requirements related to energy consumption are winter comfort, summer comfort, breathing comfort and visual comfort. In Belgium's housing facilities, there is no difficulty in the summer to avoid the use of air conditioning, because buildings can be properly ventilated and offer adequate visual comfort during the day. However, in some cases, there is no way to meet these comfort requirements, and therefore we have to use additional elements, often mechanical ones. This leads us to the second goal: the search for the best possible match between the habitat, the additional facilities and the occupant.

We suggest that architects work in this manner: to design the building that meets the occupant's goals as much as possible. If this is not enough, then we resort to technology, but not the other way around, as is often the case. When we have completed the entire design process, we will find what type of energy fits it best.

To be able to meet both goals, there are certain notions that the architect must understand, not simply know, but understand in order to be able to play with them. These notions are related to the climate and its characteristics, the comfort requirements of an individual, and the building's physical features. We pondered how to present these physical notions to architects, so that they could easily understand them and naturally integrate them into their architecture. Thus, strategies were developed, meeting the goals of the above-mentioned four types of comfort: the heating strategy (when the building needs to be heated), the cooling strategy (when it needs to be cooled), the natural lighting strategy, and the ventilation strategy. Each of these strategies is based on a number of physical concepts.

Therefore, bioclimatism means taking the climate into account from the very first design plans: considering the positive aspects of a climate on the one hand, and protecting oneself from the negative aspects, on the other.

So, bioclimatic architecture is not like the trend that is present today: stuffing buildings with more and more systems.

After 20 years, these concepts have begun to be integrated into some buildings. In this context, I ask you a series of questions:

Are we not, in a certain way, stopping architecture? We tell ourselves that we *cannot* stop architecture. Everything that is said here is at the service of architecture, in no way diminishing the architect's creativity, or the architectural quality of the architect's work. I would even say the opposite: taking these notions into account and this approach will increase the architectural quality of the project and the architect's creativity.

Is there not a risk of habitat mechanization nowadays? What are the responsibilities and the rights given to the occupant nowadays? Can the occupant still open the window? This may sound trivial, but in some recently-built buildings, they say, "above all, do not open the window".

Is there not a risk of forgetting about other consumption? We focus on winter energy consumption, but there are plenty of other types of energy involved in the building. Is there not a risk of consuming more gray energy than use energy?

1.3. Sustainable development

The end of the 1990s heralded the arrival of what is generally known as the "environmental crisis", and the "climate crisis" in particular. For those who are still skeptical, here are a few figures: in 1970, there was a concentration of 330 ppm of CO_2 in the air. At 350 ppm, climatic stability has been exceeded, and it is not certain whether we can go back. Today the level is around 410 ppm CO_2, which is huge. If nothing changes, in 2100, carbon dioxide concentrations will be between 730 and 1020 ppm, depending on the scenario adopted. The last time the 400 ppm level was exceeded was 3 to 5 million years ago: the temperature was 3–4°C higher and the sea level was 5 to 20 m higher than at present.

Architects and the world of construction are responsible for 50% of exploited natural energies, 45% of total energy consumption, 40% of generated waste and

30% of greenhouse gas emissions. We are therefore directly concerned and cannot remain indifferent to these questions.

In this environmental crisis, the buzzword "sustainable development" appeared. People used to say "I am building a passive house, so I am doing something sustainable", or "I am using ecological materials, so I am doing something sustainable". In fact, these are all varied aspects of sustainable development, but if this same passive house is 5 km from public transport and you park your 4x4 in front of your garage every day, the whole setting is not really sustainable.

Gro Harlem Brundtland, the first female Prime Minister in Europe, Norwegian and president of a commission appointed by the United Nations, concluded very quickly that the problem of sustainable development was not exclusively related to the environment. The commission quickly gave an official definition of sustainable development, which is a bit convoluted, but not bad at all. "Sustainable development is development that meets the needs of the present without compromising the ability of future generations to meet their own needs". This means that we can continue to develop, but we must change the way in which we do it.

There was also the major Rio conference which defined around 20 principles in social, political, economic and ecological dimensions. Europe has retained some of them, abandoning others which concern us less, such as the right to education (fairly acquired in Europe), or political notions (in the sense that Europe is a democracy). There are still the social, economic and ecological dimensions.

From 2000 onwards, the "Architecture and Climate" team wondered how to define sustainable development for architects. After three years of reflection, we propose:

Architecture is ethically part of its social, political, economic and environmental surroundings when it benefits from the advantages (of these four areas), while counterbalancing the inevitable disadvantages that these surroundings may present (for example, if you are in a place where there is a lot of noise, there is an acoustic disadvantage), but, above all, by letting the surroundings benefit from what we, the architects, generate. If you are planning a building that does not add anything to the environment where you are building it, do not build it. Likewise, if a building can be demolished to add value to its location, demolish it.

Therefore, architecture falls within different scales, ranging from the material to the building, to its environment, its district, its city, its territory. Talking about the building alone, like the glass tower deposited by helicopter in Sao Paolo, is over.

1.4. Smartness

In English, the term "smart" can mean "an elegant, refined person". In urban planning, the Smart City has been translated as "Intelligent City".

From the beginning, in around 2010, the "Architecture and Climate" team thought that a Smart City is not a High Tech City. Nowadays, the connection between the two is often made.

Based on this observation, we say that humankind must remain at the heart of our concerns in three fields: economy, society and the environment.

And everything High Tech-related should be at the service of these concerns.

Nowadays, there is a new buzzword appearing in the literature and in thought: quantum. Not much reflection has been given to its meaning yet.

1.5. The "Architecture and Climate" research team

All these suggested concepts have been developed by the "Architecture and Climate" research team[1] from the Faculty of Architecture, Architectural Engineering and Urban Planning of the University of Louvain-La-Neuve, in Belgium.

To develop a team, a real team, in the sense of a football or a rugby team, you have to bring people together through goals that everyone can share.

The "Architecture and Climate" team has three goals.

The first goal is to develop and disseminate scientific knowledge. As academics we have to validate what we say, we cannot just make assumptions. Our audience is architects, and our areas are: energetics, lighting, materials (from raw material to recycling and material reuse), water (which we consider to be the next big world problem after energy), waste (potentially new materials) and historical heritage. On every occasion we focused on a definition of things, which allowed us to have a vision: the "Architecture and Climate" vision. It is important to disseminate this information in publications, but also to transmit it by means of conferences, courses and meetings. It is necessary to create synergies, to develop a worldwide network of reflection, so that results are even stronger and more in-depth. We have to become involved in the public and the regional debate, to get out of the ivory tower and keep

1 Available at: https://uclouvain.be/fr/facultes/loci/archi-cli.

a realistic approach, to stay firmly anchored in the ground. We have to translate the results into a language architects can understand.

The second goal reflects our values: to cultivate a team spirit, being open to the world (to host, to travel), to ensure a work of excellence enabling each team member to flourish and develop personally, to apply the principles of sustainable development on a daily basis as much as possible.

The last goal stipulates that each research project must aim for excellence. Even if this is rarely achieved, it should be a target. It is advisable to communicate one's research, to compare results with others in order to reassure oneself and feed one's own reflection. To be a good researcher is to be creative, analytical, and above all, to be very critical of one's own work.

Our first books came out in 1983. We were very proud, because they were taken up by the European Commission and distributed across European Schools of Architecture. They were also taken up by the energy department in the United States and distributed to architects. At that point, we used to discuss winter thermal comfort, which was not new, but also summer thermal comfort, which was relatively new, and natural ventilation and natural lighting, which was completely new.

On every occasion, each research project is placed within its context, even if we are doing very specific research. It is a bit like a puzzle: cutting-edge research is one piece of the puzzle, but once the research project is completed, the result of that piece is put back into the puzzle. Here are some examples of research developed in the team: the thermoair circulation, double-skin facades, the launch of the Energie+ encyclopedia, lime hemp plaster, the urban square and the microclimate, the renovation of the built heritage with interior insulation (which is rare in Belgium), the construction of a model house in 1992 with 250 measurement points and a consumption of 18 kWh/m^2 energy need (compared to a 450 kWh/m^2 energy need at ENSAM), a study of emergency tents for the Red Cross which can respond to several climates and be light and small for transport by plane and the publication of the Monitor's "Treatise of architecture and bioclimatic urban planning" together with ADEME.

In terms of light and lighting, we created an artificial solar system for models. In terms of light, there is no scale effect. If you build a model with the colors matching your future building, you will take exactly the same photo in the model as in your future building.

Regarding water management at the district level, rainwater is not supposed to return to the sewer. It should return to the ground in order to supply groundwater, which is not obvious for increasingly mineralized urban planning. We have developed software that is based on soil permeability, the type of rain or the moment when the rain arrives. For example, when the ground is completely hardened and a strong thunderstorm arrives, water does not seep in. At that moment, it would be necessary to have very fine, typically Belgian drizzle, which sprinkles and moistures all over. The software can inform landscapers about the dimensions and types of plants for valleys.

We are working on heat storage in materials. Our dream is to find a solid-solid phase change material. When this material is found, a building will never need to be heated or cooled again. It will be a small revolution in the construction world.

In terms of education, energy managers have been trained since 1989. We developed the EDUCATE – Sustainable Architecture project at the European level, whose purpose was to compare the place of sustainable development in the curricula of European architecture schools and to suggest accompanying actions.

The activities were developed within the International Energy Agency of the European Community of the Federal State of Belgium by means of research and energy contracts, and, from time to time, in collaboration with the industry.

In 2005, we created the MATRIciel spin-off[2], because we gave a lot of advice to developers, design offices and architects, without being able to take legal responsibility. MATRIciel is a research and consulting office in energy, the environment and special techniques (as TRIBU is in France).

Here are some results in figures, provided by the research team: 3 international conferences, 25 books, the Energie+ website (Cstb in France), 115 articles over the last 10 years and 32 doctorates presented.

1.6. Teaching

In 1995, the European community was looking for a way to get the bioclimatic message across in architecture schools without having to rewrite an umpteenth book on the subject. Teachers had to be given material, and what is better than existing exemplary buildings? We presented the educational portfolio. It was a portfolio containing educational material by means of examples drawn from bioclimatic buildings in Europe. There was a portfolio on housing, on the tertiary sector, on

2 Available at: www.matriciel.be.

schools. We found all the materials that could spark the interest of any architecture teacher, whether at the workshop, for theory courses, in terms of structure or the physics of building. There was even a floppy disk with Autocad with a building model for the computer teacher. For the structure teacher there were the plans with the details in 1:10 and 1:20 scales. The portfolio also included overlays, slides, the architect's resume, sometimes an architect's text, climate and energy analysis. All the schools got one of the three portfolios, so there must be one at the school here.

Various volumes of the "Guide to bioclimatic architecture" have been published.

In 1996, the European Master in Architecture and Sustainable Development was created together with the Toulouse School of Architecture and the EPFL in Lausanne.

In 2003, we won the Roberval Prize for Higher Education with the book *Natural Lighting in Buildings*.

The educational approach to teaching in Louvain-La-Neuve is made up of three blocks:

1) three-year Baccalaureate ("*License*" in France, equivalent of a Bachelor's degree): raising students' awareness about the issues we are interested in;

2) two-year Master's degree: validation of awareness;

3) doctoral studies: reflection on this subject.

There are theory classes, project workshops and optional courses. The theory and the projects fully interact with one another. The specificity of our school was that the same teachers who taught theory lessons also conducted the workshops, which I think was specific to our school. For example, a first course on materials or climatic architecture is directly related to the first "habitat" project. The course on the physics of building (a theory course) then had an impact on the second project: natural light, with a model work in relation to the sky and the artificial sun.

These are still buildings in the city, and not in a potato field: for us, the problem of tomorrow is the city or a village (an urbanized area). We would never commission a project for a house with four exterior walls.

At the beginning, it was the results of the "Architecture and Climate" course that had an impact on the other courses. Afterwards, in the field of research, it was the researchers of "Architecture and Climate" team who became involved.

Optional courses are those which can be found at other faculties, such as for example, on environmental pollution, environmental law or ecology. During the last year, there is the graduation or final thesis (the *"dissertation"* in France), which is usually an architectural project.

1.7. Conclusion

As a conclusion, I would like to present three questions.

The challenges of tomorrow are multiple: the territorial and programmatic aspect, air quality, the environmental quality of materials, waste, water issues, biodiversity, etc. We should not legislate as we have legislated so far in relation to energy. This is the great drama of environmentalists, who have tried to spur change with simple speech, but who have seen nothing concrete. It then became necessary to legislate, but we have undoubtedly legislated too much, too much has been cut from the architect's creativity.

The passive has two dramas: firstly, in passive mode, we cannot be autonomous in terms of heating and secondly, there has been an increase in the amount of materials (insulation and additional sheaths, the increase in the volume built), which totally goes against the idea of economy of materials.

The third point to which I hold: we must not forget the human being. Humans must keep control of their habitat and all the systems in which they find themselves immersed.

So, to answer the question asked: obviously the architect must be part of the solution; the architect can in no way be part of the problem!

Thank you for your attention.

1.8. Questions

1) Student at ENSA Marseille:

What struck me in the presentation was the use of the term sustainable development rather than sustainability. In sustainable development there is the allusion to development and durability which confronts the building with its finiteness: we want it to stay forever. Are the two terms the same for you?

André De Herde: We looked for another term for a very long time. We found no other better term than sustainable development. But it is true that we should talk about architecture as something that needs to be supported (the French term "*soutenir*" can both be translated as "sustain" and "support"), in the same way that a supporter supports a sports team, but the expression "sustainable architecture" did not work. "Sustainable development" was the expression used by everyone, which is why we made a definition that is different from the definition of something "durable over time", like concrete, for example. But you are right, it is "sustainable development" in the sense that it can be supported, not in the sense "that it can last over time". However, if you find another word, I am open to suggestions.

Reactions in whispered words come up from the audience and we hear the word "growth".

André De Herde continues: I think that Brundtland is right, we must be able to continue to develop, but in what ways and how? It is not the development of multinational companies; it is joint development, of value, of more natural elements, of low technologies and not of high technologies. This is the development of the future, for me.

2) Student at ENSA Montpellier:

You used a phrase that impressed me: "If you are planning a building that does not add anything to its environment, then do not build it. On the contrary, if demolishing a building adds value to its environment, then destroy it." Yesterday evening I was at Philippe Madec's conference, which presented a rehabilitation of the Multimedia library and the refugee house. His speech was different: we can no longer afford to destroy a building, because today we build too much, we use too many materials. Demolishing the Multimedia library building could have helped the environment, and so could rebuilding it with a new building, but he decided to keep the building. So where does rehabilitation fit in your speech?

André De Herde: I am convinced that there is no contradiction. I think Philippe Madec wondered whether the building should be demolished or kept. If he estimated *in fine* that it was worthwhile to keep it, I think he made the right decision. You do not have to demolish the building every time, but sometimes this makes it possible to create a public space, for example, which can provide a benefit. If Philippe Madec had felt he had to demolish the building to do better, he would have done so.

Robert Célaire: Buildings must not be demolished; they must be deconstructed. We have to revive them and reuse the materials.

De Herde: Yes. MATRIciel is currently working on a building without any new material brought to the site, everything is recovered or recycled.

3) Teacher in Marseille:

When you said that we should invent a material that releases and absorbs heat slowly, this has already existed for a long time: it is raw earth.

André De Herde: In Belgium, if you use raw earth, you will have to heat the building at some point, unfortunately, unless you accept a temperature of 16°C inside. But lime hemp plaster has extraordinary qualities.

Robert Célaire: André was referring to the latent heat during phase change, that is, materials that store energy at a constant temperature. It is a research theme.

4) André De Herde's former graduate:

I graduated under Mr. De Herde 20 years ago. You have in front of you an exceptional person. I would like to pay tribute to him. I came to this conference to talk about him. I am here because you are the answer to the problem. You are all future architects. I have been fighting alongside Mr. De Herde for 20 years to impose this bioclimatic, to take the environment into account in architecture. This is a very difficult thing to do every day a an architect. On some projects, there is even talk of the architectural "option". So, on the question "do architects want to be part of the problem or the solution?" I think that with this bioclimatic treatise, the architects have the solution. You all have the solution. With your BAC+6 brains, you are all able to tell everyone "we know how to manage the environment, with technical and architectural answers". I would like to ask all of you to fight to make the profession of architects last. I would like the flaming torch to be passed on, I was one of his disciples and I would like there to be more followers. I ask you all to invest in it and believe in it, to drop all these pessimistic speeches that say that it is the end of the world. It may not be easy, but we have to keep some optimism.

The audience greatly applauded this speech.

5) Former student of ENSA Montpellier and the Réunion branch:

I am referring to the first question, and the term sustainable development. It has been said that it cannot be replaced by the word sustainable. Can we not replace it with the term "low tech"? I saw a lot of buildings which were passive, but which were not necessarily low tech or sustainable development-oriented. On the other hand, all the low tech buildings that I was able to study correspond to the criteria of sustainable development architecture.

André De Herde: That is a good suggestion. I am afraid we might stay too attached to "tech". Sustainable development architecture also has a lot to do with the program, for example, and the situation in the city. And then, we call a table a table, and a chair a chair. So if the word consecrated by the public is "sustainable development", why not use the word sustainable? But you have to define it. If you define "low tech" as I define sustainable, that is all right and we agree.

2

Architecture, City, Landscape: The Place of History in Pedagogy Integrating the Climatic Emergency

2.1. Introduction

This chapter presents a reproduction of a roundtable discussion dedicated to the historical dimension of climate issues. It demonstrates how history makes it possible for other disciplines taught in architecture and landscape schools to be put into perspective. This question is addressed through three inputs: the landscape approach to climatic heritage, the challenges of teaching environmental history and the integration of climate history into the teaching of architectural history. This involves conducting an inventory and reflection on the evolution of content, bibliographies, exercises and methods for evaluating these lessons, highlighting the links between research on climate change (particularly in its historical dimension) and training in the field of design disciplines.

> L'homme par son égoïsme trop peu clairvoyant pour ses propres intérêts, par son penchant à jouir de tout ce qui est à sa disposition, en un mot, par son insouciance pour l'avenir et pour ses semblables, semble travailler à l'anéantissement de ses moyens de conservation et à la destruction même de sa propre espèce. En détruisant partout les grands végétaux qui protégeaient le sol, pour des objets qui satisfont son avidité du moment, il amène rapidement à la stérilité ce sol qu'il habite, donne lieu au tarissement des sources, en écarte les animaux

Chapter written by Anne COSTE, Frédéric DELLINGER and Theodore GUUINIC.

For a color version of all the figures in this chapter, see www.iste.co.uk/aithaddou/transition.zip.

> qui y trouvaient leur subsistance, et fait que de grandes parties du globe, autrefois très fertiles et très peuplées à tous égards sont maintenant nues et stériles, inhabitables et désertes. Négligeant toujours les conseils de l'expérience, pour s'abandonner à ses passions, il est perpétuellement en guerre avec ses semblables, et les détruit de toutes parts et sous tous prétextes : en sorte qu'on voit des populations, autrefois considérables, s'appauvrir de plus en plus. On dirait que l'homme est destiné à s'exterminer lui-même après avoir rendu le globe inhabitable. (Lamarck 1830, pp. 154–155)
>
> [Due to his selfishness, Man is too little clairvoyant for his own interests, due to his inclination to enjoy everything that is at his disposal, in a word, due to his carelessness about the future and for his fellow humans, he seems to be working towards the annihilation of his means of conservation and the very destruction of his own species. By destroying at large the plants protecting the ground, for the sake of objects to satisfy his momentary greed, he quickly brings to sterility this ground which he inhabits, gives rise to the drying up of sources, removes the animals which found their subsistence there, and causes large parts of the globe, once very fertile and heavily populated in all respects, to presently be bare and barren, uninhabitable and deserted. Always neglecting the advice of experience, abandoning himself to his passions, he is perpetually at war with his fellows, and destroys them everywhere and under all pretexts: in such a way that we see once formerly considerable populations, increasingly become impoverished. It seems that man is destined to exterminate himself after making the globe uninhabitable.]

Awareness of the devastating effects of humans on the planetary ecosystem is not new, as evidenced by this two-century-old text by French naturalist Jean-Baptiste de Lamarck[1]. Generally centered on the fruits of human genius, as shown in the image of Marrakesh (Figure 2.1), the history of architecture, the city and the landscape also provides rich testimony on the genealogy of the crises that we are experiencing.

How has consideration of climate change altered the teaching of architectural history, urban planning and landscape? How does the historical dimension make it possible to put into perspective the climatic questions addressed by other disciplines

---

1 Excerpt that we owe to the investigations of Nicolas Vernet, doctoral student at the AE&CC research unit in Grenoble, as part of his thesis, "*Garden cities* for the 21st century: The garden city model as one of the possible responses to the housing crisis in a context of energy transition?" (AE&CC/ADEME, direction A. Coste, co-supervision S. Sadoux).

taught at ENSA/ENSP (*École Nationale Supérieure d'Architecture* – French National School of Architecture/*École Nationale Supérieure de Paysage* – French National School of Landscape Architecture)? These are the questions to which we have collectively tried to provide preliminary answers, based on our various experiences.

Figure 2.1. *View of one of the courtyards of El Badi Palace in Marrakesh, 16th century (photo F. Dellinger)*

Coming from different backgrounds, generations and disciplines, we intend to reflect on the basis of three clearly identified inputs: the approach to climatic heritage through the landscape (here, the garden), the challenges of teaching environmental history and the integration of climate history into the teaching of architectural history.

Taking into account the history of the climate and, more broadly, that of the planet, requires reflection on temporalities and time scales, in the sense of a mental zoom-in/zoom-out exercise which completely modifies the relationship to the usual objects of the history of architecture, generally western-centered but also *Homo sapiens*-centered, so to speak. This often begins with the first architectural testimonies of a form of domination of nature by humans, while at the same time retaining some of the humility of these societies vis-à-vis the natural phenomena (the site of Stonehenge is one of these ambivalent testimonies, for example). At that time, Man was living under the gaze of the gods. It is significant that the oldest achievements, reduced to a small if not long excluded portion of history books and school curricula, are those that precisely highlight the relationship between architecture and cosmology[2]. In the same vein, at this time of profound crisis, we must question the absence in our narrative of the most sober and most resilient examples of architectural production, fruits of the culture of people most in harmony with nature and the Earth. While the emphasis has been put on the technology, the power, the dominance and the spectacular dimension of architecture in history teaching – from the history of cities to the history of buildings – research that might once have been considered insignificant, such as that on earthen architecture, is now a great resource for rethinking the terms of our teaching. In a similar way, while the camp of capitalism, technology and the excessive consumption of matter and energy has become entrenched over the last 200 years, the history of its opponents – from the utopias of William Morris (1890) to Rob Hopkins'[3] cities in transition – constitutes a whole field of study likely to shed light on the issue at hand.

The integration of the climate thus invites us to re-examine the history of architecture at its foundations and to identify the frames of reference it offers in order to grasp the current issues. How did the societies of the past, which had/took time for empiricism, come to own their territory? The concept of precedent in terms of adaptation to resources and to climate, but also in terms of crises, can be used in the teachings of ENSA and ENSP. Let us observe that at ENSP the acknowledgment of the passing time as a preliminary consideration, as well as the interweaving of geographical scales, are key issues in the teaching of the design process. The landscape project is therefore a bridge between an assumed history and a desirable future for the area of intervention. As these temporal scales are approached integrating new dimensions – history of the Earth, history of climate, history of energy – evolution elicits very different perspectives from previous practices.

Based on the different experiences of teachers, the roundtable that brought us together during the EnsaÉco meetings at Montpellier intended to take stock of the evolution of content, bibliographies, exercises and evaluation methods for these teachings, both on the questions that these changes raise for teachers and on open questions. Also intended to highlight the links between research on climate change (particularly in its historical dimension) and training in the field of teaching the history of architecture, this roundtable was considered the starting point for a longer-term work exchange of experiences and reflections on this theme, channeled through the EnsaÉco network.

---

2 Let us mention the ongoing doctoral work of Anaïs Guéguen-Perrin: Guéguen-Perrin, A. (N/A). BuenVivir and Habitat, Environmental and cultural questions in indigenous territories and habitats in Brazil and alternative ones in France. Thesis supervised by Thierry Joffroy within the AE&CC research unit, in co-supervision with the Habis IAU/USP of São Carlos.

3 Initiator of the now global movement of cities in transition, Rob Hopkins is the author of several books that inspire our teachings, including Hopkins (2008, 2019).

This contribution combines the testimonies of the three authors, Frédéric Dellinger for the first section (landscape dimension of climatic heritage), Théodore Guuinic for the second section (for an environmental history) and Anne Coste for the last section (testimony of the evolution of his teaching in Grenoble), enriched by the exchanges they could benefit from at ENSA Montpellier. It concludes with avenues the authors wish to discuss within the architecture and landscape education network.

2.2. Climatic heritage through the landscape

The question proposed in this section mainly concerns the solutions within our reach to the heatwaves present and yet to come in France and in Europe, via the study of climatic heritage and the concept of "landscape". By "climatic heritage" we mean heritage developments and constructions, evidence of a process integrating the strengths and limitations of a climate when occupying a territory.

2.2.1. *A general trend but individualized local situations*

Each year, scientists' predictions of potential changes to the planet's climate are being refined. This is particularly true of changes to come at the local level, namely the climatic disruption which will affect us all in the future. Lyon could have Madrid's climate by 2050 and that of Algiers by 2100[4].

Every summer, we experience heatwaves in our towns, cities and homes. Today we have the technology, a whole range of high-tech solutions to ensure thermal comfort in summer and to regulate this comfort remotely (an important issue of the smart city). Not only do these technologies often consume energy, but they also tend to worsen the heat islands they are supposed to combat. So we cannot continue like this. It is up to architects, urban planners, landscapers and engineers to participate and propose "new" ways to support our hometowns and beyond in developing resilience against this disruption. How can the history of the city, the gardens and the landscapes lead us to these new perspectives?

One of the main changes brought about by global warming is the modification of the local climate. We are already seeing an increase in the intensity and frequency of

4 Working hypothesis adopted by the Métropole de Lyon for its "climate and territorial energy plan", in its brochure Adaptation to Climate Change – Base of Local Knowledge, available for consultation and download on the Métropole website: https://blogs.grandlyon.com/plan-climat/.

climatic phenomena (rain, wind, snow, heatwaves, etc.), which is expected to further increase due to warming. Specifically, we can interpret it as an extension/translation of climatic zones, as described by the German climatologist and botanist Wladimir Peter Köppen at the beginning of the 20th century (Figure 2.2). These climatic zones are characterized by temperature and precipitation, combined with plant associations. For Europe in general and France in particular, this translates into an extension of the Mediterranean climate to the north and an increase in climatic levels in mountainous environments.

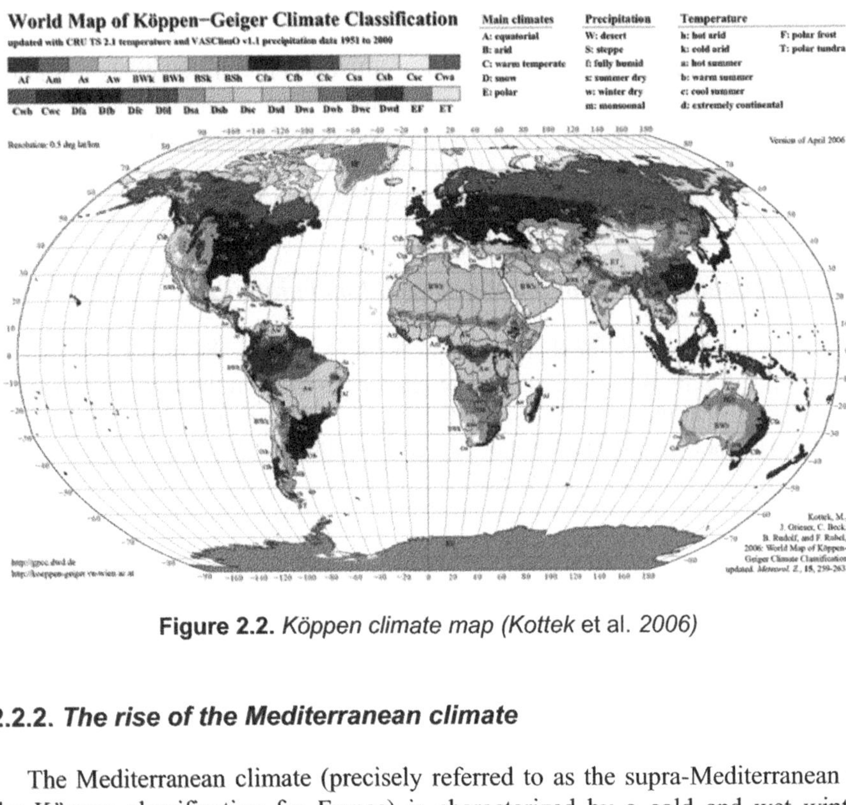

Figure 2.2. *Köppen climate map (Kottek* et al. *2006)*

2.2.2. *The rise of the Mediterranean climate*

The Mediterranean climate (precisely referred to as the supra-Mediterranean in the Köppen classification for France) is characterized by a cold and wet winter season (with a high risk of flooding) and hot and dry summers (with high risk of droughts, heat waves and fires). Let us note that the Mediterranean climate is itself changing to become more and more like the desert climate on its southern borders, in North Africa.

Neither the fauna (and even less the flora), nor our habitat, are prepared to cope with these climate changes so quickly.

However, the history of civilizations teaches us that, thousands of years ago, humans were able to conquer almost all environments, from the hottest to the coldest. This is illustrated by Figure 2.3 which shows villages in the South Atlas in Morocco, built in stone and earth with flat roofs and small openings, adapted to the harsh winters and to the heat of the Moroccan summer. Humans knew how to implement simple, low-tech, low-resource solutions which are still effective today. How can the analysis of heritage, of constructive cultures – the understanding of the physical and agronomic phenomena implemented and used by our ancestors – still be a source of hope and opportunity to improve our comfort and thus our daily life in our neighborhoods and homes?

Figure 2.3. *South Atlas village in Morocco (photo: J. Lelay)*

2.2.3. *What models are there in this context?*

El Badi Palace in Marrakesh and its garden courtyards (Figure 2.1) were built in the 16th century. The layout is both emblematic of the art of Arab-Andalusian gardens but also, in the Mediterranean world, of palaces and houses in relation to the street (evolution of the patio house from Antiquity), to the courtyard and the garden. This set is particularly rich in lessons on the climatic comfort of housing in Morocco's semi-arid climate, but also on the development and optimization of local resources, such as land, water and plants. Beyond the issues of approval and representation of the power of its sponsor, the Saadian Sultan Ahmed al-Mansur ad-Dhahbî, this courtyard-garden offers an oasis of freshness at the heart of the palace, as well as a citrus garden. The writings and testimonies about the palaces of this period also evoke olive trees, date palms, pomegranate trees, Argan trees, as well as many ornamental (the famous rosebushes of Damascus) and medicinal plants. This fits quite well with our definition of urban agriculture today.

The Arab-Andalusian garden constitutes a technical and agronomic apogee of the art of Persian gardens, both in its effectiveness on the preservation, enhancement and presentation of the "water" resource, as well as for improving plant species and cultural practices. In his book *Les jardins : paysagistes, jardiniers, poetes* (1998, p. 209), Michel Baridon describes it as: the "Garden of the planet's dry belt. Son of the desert, it is the oases that it first brings to mind". In this art of gardens, we find all the efficiency of fruit and vegetable production in the oases, namely thanks to the intelligent management of water through irrigation and working on fertile soils. The formalism of the garden, imbued with symbolism, esotericism and aesthetics, appropriates cultural techniques that are probably much older. Within these residences, the evapotranspiration of plants, maintained by means of irrigation, constitutes the main source of "air conditioning" for the inhabitants. This active air conditioning complements the "passive" temperature conditioning of buildings. The "passive" techniques, among other things, make use of the properties of earthen constructions (for this palace in Marrakech), the albedo of plaster, the properties of the air currents induced by the geometry of cities and buildings, the optimization of openings or the use of shade houses and other *moucharabieh*, to name but a few examples.

2.2.4. *Lessons for teaching*

At that time, architecture, urban planning and the landscape were closely related, inseparable disciplines, in order to meet the challenges of quality and comfort expected by inhabitants. What importance should be given to teaching the history of gardens, landscapes and the city, in the training of architecture and landscape students? How can understanding the actions implemented by these civilizations to adapt to their environment be a source of inspiration for the generations to come? While our ancestors probably relied heavily on empiricism to adapt, how can the technical and scientific analysis of the surroundings of cities, districts and buildings still present be a training medium for reflecting about the city of tomorrow? Past generations took (Did they have a choice?) the time to try things out through experience. How can this knowledge be valued today? Would it be possible to appropriate their building culture – at least in part and not merely from a formal point of view – to accompany this "rise" of the Mediterranean climate in Northern Europe?

We are only at the beginning of the disruption resulting from global warming and the answers will not all be drawn from technological innovations. The teaching of ecological transition will also involve valuing the technical and empirical knowledge of the generations that precede us in neighboring geographical contexts. On the basis of our ancestors, perhaps it will also be necessary to (re)learn the humility of experimentation and developing long-term projects, without being able to control everything. This will require not only taking the risk of making mistakes, but also accepting the fact that we will not be able to witness the result of all our work nor the accuracy of our assumptions during our lifetime.

One of the first levers for teaching in this context requires us to modify our relationship with time: with past time, with the lessons from history; with the future, involving a project time span exceeding a human lifetime, (the) time dimension of landscape and nature. The following two sections each analyze these time-related questions in their own way: first, by examining the challenges of teaching environmental history in Architecture training, and finally showing how the results of recent research works are changing history teaching on the one hand, and providing some prospective tools on the other hand.

2.3. What lessons can be learned from environmental history?

While it is true that for about 15 years, an "education for sustainable development" has been implemented in primary and secondary education (Considere and Tutiaux-Guillon 2013, pp. 111–133), baccalaureate holders who enter a school of architecture do not yet seem to have taken on board the full measure of the environmental catastrophe. The lessons received about this topic in the context of higher education are decisive not only for their professional, but for their personal and civic training. *A fortiori*, this would be applicable to future architects, in a context where the prescribing architect has often remained the accomplice of a highly polluting and carbon-intensive sector of the economy (Ménard 2014, pp. 161–168). Despite having been endorsed 10 years ago[5], the desire to intensify training in environmental issues in higher education still has a long way to go: as shown by the first national inventory (Berthault *et al.* 2019), for the training courses recently conducted, consciousness awareness about environmental and climate issues still affects a minority of higher education[6].

5 Grenelle I law, Article 55 (2009).

6 The study estimates that only 11% of training courses include compulsory courses on the energy–climate issue, mainly at Master's degree level, whereas courses relating to the environment remain optional for the most part, and are mainly provided within specialized courses.

In this context, and regardless of the training courses implemented, there is a new generation of students who no longer hesitates to express their expectations, as can be evidenced by several forums and recent actions[7]. Encouraging the support of many school heads and teachers[8], these actions contributed to bringing to the National Assembly a bill aimed at generalizing the lessons related to environmental and climate issues[9]. Forging its way through overabundant and inconsistent information, in recent years this generation has been able to find additional insights among popularizers from the scientific world[10], providing them with a better understanding of the environmental disaster, and raising more and more consciousness in them. Today, this concern has even been transmitted to high school students who, inspired by the charismatic figure of Greta Thunberg, are now expressing increasingly precocious demands. Thus, we can bet that hereafter, on the threshold of their higher education, future students of architecture schools will share a growing and unanimous interest in the ecological question. It is in this context that it is advisable to question the evolving methods of the courses offered as a whole, without limiting oneself to the courses already devoted to environmental problems. So what role can history assume within this new framework?

Due to its unprecedented scale, the current disruption constitutes a prominent historical event at the scale of the history of the planet, and cannot leave unchanged the content and vocations of history teaching provided to architects. An examination is therefore necessary and prompts us to consider what narratives might give meaning to the state of the world that our students will inherit, and within which they will have to exercise the profession: What history(ies) should we teach them for them to learn how to build in a ruined world?

---

7 Let us quote here the recent "Student manifesto for an ecological awakening" launched on September 25, 2018, following the initiative of students at the École Polytechnique, as well as the Global student manifestation for the climate on March 15, 2019.

8 Collective opinion piece published in the *Journal du dimanche* on September 14, 2019.

9 Draft law of September 25, 2019 on "the generalization of the teaching of issues related to the preservation of the environment and biological diversity and climate change within the framework of planetary boundaries".

10 By way of example, we can quote here the works of Jean-Marc Jancovici or Pablo Servigne, in their respective vocations.

2.3.1. *Synthesis lessons and historiographic points of reference*

Based on repeated exchanges with undergraduate students[11], the latter sometimes present very unequal levels of understanding of ecological issues. So, as a prerequisite to reduce these disparities, it might be useful to impart general lessons that succinctly explore the knowledge whose scope can be discouraging, despite the multiplication of available resources[12]. The all-encompassing point of view specific to history would make it possible to retrace environmental issues along the vast period of time when they unfolded, set against the framework of a history of climate and the environment. At the same time, and in synergy with these teachings, a history of techno-critical thought and great thinkers of ecology could also be part of the Theories and Doctrines course[13], making it possible, for example, to clarify shared concepts such as "sustainable development" or "environmental crisis". Unlike the context of high school education, the challenge in architecture school is not so much about delivering a broad introduction, but instead enriching the core of already acquired knowledge with a reflexive and critical dimension. This is why history might move from a mere observational attitude to shed light on the causes, and provide an insight into the multiple factors that cause environmental disruption, the comprehension of which is truly essential.

2.3.2. *Environmental history as a horizon or as a viewpoint*

Responding to such an ambition would also lead to relying on recent achievements in a cultural history of the environment (Mathis and Pépy 2017), that of the history of techniques (Edgerton 2006), in particular energies (Auzanneau 2015; Massard-Guilbard and Mathis 2019) and technological risk (Fressoz 2012): in a word, to recompose the picture of an environmental history[14] that adjusts to the needs of architecture education (Figure 2.4).

---

11 In line with the courses provided, on the basis of a survey carried out in spring 2019 in the form of an online questionnaire sent to a group of students from the Montpellier National School of Architecture (Bachelor's degree 2, semester 4).

12 Regarding online resources of university origin, we can quote the Virtual University for the Environment and Sustainable Development (UVED), and the *Moocs* offered by various universities, especially American ones.

13 This is the approach of our course "Architecture in times of crisis", Theories and Doctrines, CM, S4, ENSA Montpellier, adjunct lecturer T. Guuinic.

14 In the following lines, we rely on the "Historiographical assessment of environmental history", presented by Geneviève Massard-Guilbaud to the General Assembly of the Association of Contemporary Historians of Higher Education and Research (AHCESR) on November 6, 2010. Available at: https://ahcesr.hypotheses.org/ [Accessed November 8, 2019]. On the subject, see also Quenet (2014).

Figure 2.4. *View of the beach and the oil fields from the pier at Huntington Beach (California). Anonymous photo circa 1940, Orange County Archives*

Born in the United States in the late 1960s[15], this approach has experienced a delayed boom in France, showing increasing popularity in the past 20 years. Placed in the wake of a history of the environment, it nonetheless differentiates itself by focusing its attention on the relations between humans and the environment, in its material, cultural or even political components. An intrinsically multidisciplinary discipline, environmental history nourishes itself as much from the achievements of the history of science and technology, as from cultural history. This is why it seems to be less of a new sub-discipline rather than to proceed from a broadening perspective and the ambition of a synthesis of knowledge spanning across the scale of several centuries.

2.3.3. *Environmental and long-term history*

Environmental history is thus different to the history of the environment which, fueled directly by the increase in scientific data, was long the work of researchers outside the historian's profession. Faced with this observation, and the lack of effect of the alerts launched by climatologists, a manifesto (Armitage and Guldi 2014) published in 2014 called on historians to become involved in the public debate. Invoking the opportunity offered by the massification of data (Big Data), the work took the form of a plea in favor of studies carried out over a long period, in the distant wake of the reflections of Fernand Braudel. The authors pleaded for a unifying history of knowledge, more attentive to contemporary issues, and taking as an example the case of environmental history. Beyond the controversies provoked by the book, the debate initiated had the advantage of making people think about the benefits and methodological difficulties raised by this new scale of reflection and the scopic games to which it invites. It is in this context that Carlo Ginzburg recently proposed the idea of analyzing "*longue durée* (historical phenomena), in detail"[16]. While such reflections still need to be deepened, the widening of the field offered by the "*longue durée*" could already favor the understanding of a contemporary era conceived as "endless crisis" (Revault d'Allonnes 2012).

2.3.4. *Critique of concepts, alternative narratives, a space for discussion*

Under the effect of a similar jump in scale, Jean-Baptiste Fressoz and Christophe Bonneuil proposed to consider the Anthropocene less as a period than as an event (Bonneuil and Fressoz 2016), thereby inviting us to identify its roots. Recalling that

15 The emergence of the discipline is often associated with the publication of Nash (1967).

16 Carlo Ginzburg's "*La longue durée, à la loupe*", lectures given from May 4–26, 2015 as part of Roger Chartier's course at the Collège de France.

to measure is not to understand, the two historians based their arguments – albeit controversial ones – on multiple works on the history of techniques and social history, in order to specifically refute the idea of a newfangled awakening of environmental awareness or concern about technological risks. In its singular vocation, this essay illustrates the critical function that environmental humanities can exert, by challenging the relevance of generally accepted notions, such as sustainable development, or, in this case, the Anthropocene. In the light of this example, one can imagine how the lessons in environmental history can contribute to critically apprehend the concepts of ecological transition or collapse, which are highly present in the public debate today.

By deploying explanatory accounts of climatic upheavals, environmental history might finally have the advantage of offering a space for discussion open to the interpretation of the causes that have led to the current situation. Beyond its political and civic function, this set of narratives could potentially come to be owned by young students, and be more stimulating than a unique, hegemonic vision, with *a fortiori*, no exit. Accepting to act as a relay, then, would not be limited to responding to the concerns of students by simply agreeing to be the bearers of "bad news", but might involve thinking about the frameworks and speeches best suited to match student sensitivity.

2.3.5. *Environmental history: spur or model?*

For all these reasons, environmental history offers itself as a promising resource to support students in understanding an environmental disaster which, apart from their professional future, affects their very condition. Now, if such insights can contribute to renewing the teaching of history in schools of architecture, to what extent do they constitute a novelty? On the one hand, these teachings would follow in the wake of the first educational initiatives[17] and the research dynamic engaged in the field of a history of architecture read through the prism of energy[18]. On the other hand, and in a more general way, they would also be part of an already initiated expanding movement, embracing a plural teaching of history which acknowledges several voices in the schools of architecture, focusing on a history of architecture which has long been familiar with the crossbreeding of art history, the history of techniques, that of architectural theories or the history of the cities.

---

17 We can quote here the teachings of Sébastien Marot at the École d'architecture de la ville & des territories of Marne la Vallée (Paris-Est).

18 Here we are particularly thinking about the research works of Emmanuelle Gallo and, more generally, the national research program, *Ignis Mutat Res*, given at the Ministry of Culture and Communication between 2012–2015.

In no way do the remarks proposed here intend to draw up an inventory of the richness of such a broad and dynamic field of research. They attempt to glimpse its possible contributions within the framework of historical practices in schools of architecture, which means there are many perspectives to be explored collectively. A stimulating invitation to open up new educational perspectives[19] and research avenues (Ménard 2018), the development of an environmental history would not, however, be without methodological difficulties. These would arise both for teachers and students alike, over whom they could impose "great differences" at the thematic, chronological and disciplinary levels, which would therefore require great versatility to reckon with the different approaches.

The third and last section of this chapter will introduce some tools that can help to confront some of these difficulties, as they testify to the first developments at different levels of education in the context of another ENSA institution: that of Grenoble.

2.4. Integration of the climate dimension in the architectural history course

Here we have the example of ENSA Grenoble where the reconstruction of the history of architecture course through the prism of climate issues is based on the results of research work, in particular developed by the Architecture Environment and Constructive Cultures research unit (LabEx AE&CC). As such, the "TerrHab" project (ANR 2010–2014), on the concept of habitability (Figure 2.5), and the "Spatializing energy transition" project (*Ignis Mutat Res* 2013–2015), on the concept of the territorial energy ecosystem (Figure 2.6), are at the origin of a significant evolution, not only in the content of the Master's seminars, but also in the content of the bachelor's degree lectures. Following this last project, which gave rise to various publications (for example, Coste *et al.* 2018), several doctoral theses involving a historical approach have emerged. The latter contribute to the evolution of course content. This is why, in this chapter, we will begin by presenting the research and monitoring part of our activity, before addressing the lessons of the initial course and the questions they raise.

19 It is from this perspective that we have created "Architecture and energy transition" – Specialized Master's in contemporary architecture and heritage and sustainable urban project management. Adjunct lecturer Th. Guuinic.

Figure 2.5. *Cover of the book* Les sans mots de l'habitabilité et de la territorialité, *the result of a 2010 ANR project: TerrHab –From habitability to territoriality (and back): Of urban peripheries, individuals and collective bodies in interaction*

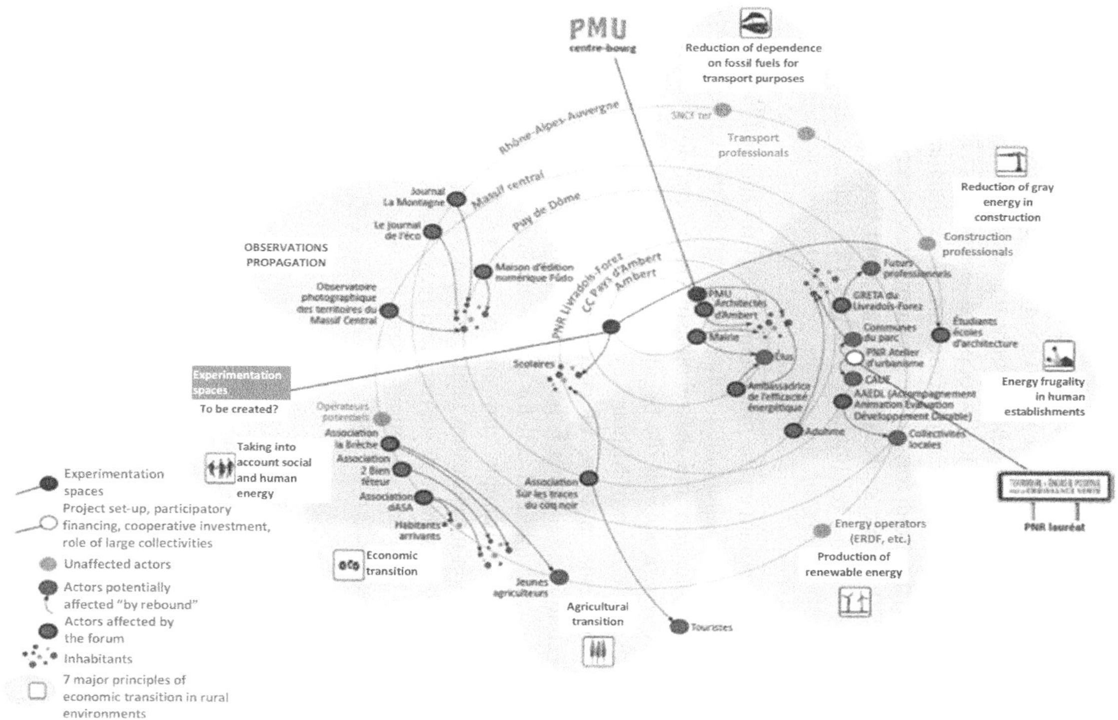

Figure 2.6. *Modeling of key players in relation to the energy transition levers identified (source: http://www.bluehouse.fr/JB6/territoires/)*

2.4.1. *With doctoral students: work on precedents*

One of our lines of research focuses on precedents in the area of crises – environmental, housing, health – and in architectural and urban theories. Our recent work has brought us to the English case of *garden cities*, in particular, and the Victorian context of political debate in the second half of the 19th century. It is within this framework that we carried out research on the technical, economic and political history of energy, bringing to light the fight that was then being waged in England between projects of diametrically opposed companies. The capitalist economic model based on the reckless exploitation of fossil fuels and the liberal ideology having prepared the field for climatic catastrophes, emerged victorious from this confrontation. Different sources show that, from that time, observers questioned themselves about the relationship between CO_2 release and climate change.

As part of this research program on the precedents of experimentation in the field of low-energy housing and/or integrating key questions in the current crisis concerning agricultural production and the relationship to nature, doctoral theses focus on *Garden Cities* experiences (at the beginning of the 20th century)[20] or the ecological habitat of the *New Town* of Milton Keynes (1960s and 1970s)[21]. The aim here is to identify precedents in terms of crises (what is comparable between England in the 19th century and today it is not the nature of the crises themselves, but the fact that they are multiple and call for ecosystem approaches), in terms of experimental methods (for example low-energy and/or low-cost housing competitions), anti-speculation measures on land, etc. In recent years, other works, focused on approaches to design through sobriety and reuse, particularly in Brazil[22] and in Europe[23], have also been an opportunity to put the issues related to climate change into historical perspective.

20 See Vernet and Coste (2017). Also: "Garden cities for the 21st century: The garden city model as one of the possible responses to the housing crisis in a context of energy transition?" (Thesis in progress).

21 See the current thesis by Malaury Forget, "Evolution of the energy frugality paradigm. The experiments of Milton Keynes' New Town in terms of habitat" (1967–2017), published by LabEx AE&CC and under the supervision of G. Novarina and the co-supervision of S. Sadoux.

22 PhD thesis by Thiago Lopes Ferreira, "Vernacular architectures and contemporary production processes: Training, experimentation and construction in a rural community in Brazil", supervised by A. Coste and A. Ino (defended at ENSAG in 2014).

23 Thesis by Marie de Guillebon, "Towards a practice of reuse in architecture. Experiments, tools, approaches", supervised by A. Coste (defended at ENSAG in 2019).

Research contracts and thesis work – combined, of course, with the resources provided by the many current scientific publications and their popularization – provide material not only for reflection, but for enriching the content of the courses. Beyond this, it is also through the exercises related to these teachings that educational changes are measured.

2.4.2. *Bachelor's degree: first and foremost, establish a relationship*

The major change in recent years in the course of the first semester of the Bachelor's degree[24] is the integration of different time scales, not only to present climatic issues and socio-ecological transitions, but also to perceive the relativity of the temporal dimension of the history of architecture in relation to the Sapiens, and that of the planet. This makes it possible to bring different temporalities forward (age of the Earth/history of hominids/history of the Sapiens/history of the Anthropocene) by measuring the respective weight of the different periods – in particular the insignificance of the proportion of time since the start of the Anthropocene – and the dizzying acceleration of climatic disorders, the destruction of biodiversity and resource consumption (McNeill 2001). In this process, works such as those by Harari (2015), Illich (1974) or Gras (2007) are highly valuable.

The second evolution is the integration of the history of energy, in terms of its relationship to architecture. The history of architecture through the history of construction, through constructive cultures and the history of materials, has always been based on work in the history of science. Today, the history of energy, the exploitation of resources and energy policies enriches this approach (Stein 1977).

It is thus easier for students to understand the role and responsibility of so-called "modern" building methods in current crises. Beyond that, it is a question of architecture "committed to the project of world habitability" (Findeli and Coste 2007) which is proposed. Many disciplines, including art – in a privileged position (Figure 2.7) – are mobilized to stimulate reflections and exchanges.

On the scale of the history of architecture, the general introduction proposes to observe the architectural and urban production of the different eras in terms of their relationship to technique and their relationship to nature. This great introduction now includes the history of energy transition, in the sense of a change in the dominant system (wood, hydraulic, coal, oil, etc.) relating to the rise in power and dominance of the capitalist system worldwide.

24 S1: "Introduction to architectural history and theory: The project in relation to the world" (Prof. Anne Coste).

Figure 2.7. *View of the work of artist Tadao Ando, "Four Cubes To Contemplate Our Environment", Château Lacoste, France, 2008–2011 (photo: A. Coste, 2019)*

In line with this multi-scalar (spatial scales, time scales) and multidisciplinary approach, the exercise carried out in parallel with the course has also evolved. For the past two years, S1 students have been invited, from the first session, to develop a chrono-systemic timeline (see Bergeret *et al.* 2015). A great deal of work is required from the students to link different fields and events and then describe these links and (Figure 2.8). This is not about developing a deterministic approach to the history of architecture; it is a way of fostering an appraoch that is both cultural and holistic. This activity, carried out throughout the various sessions, has the benefit of making the students actors in their own teaching process. Furthermore, it trains them to think in an ecosystemic way, rather than a linear or compartmentalized one.

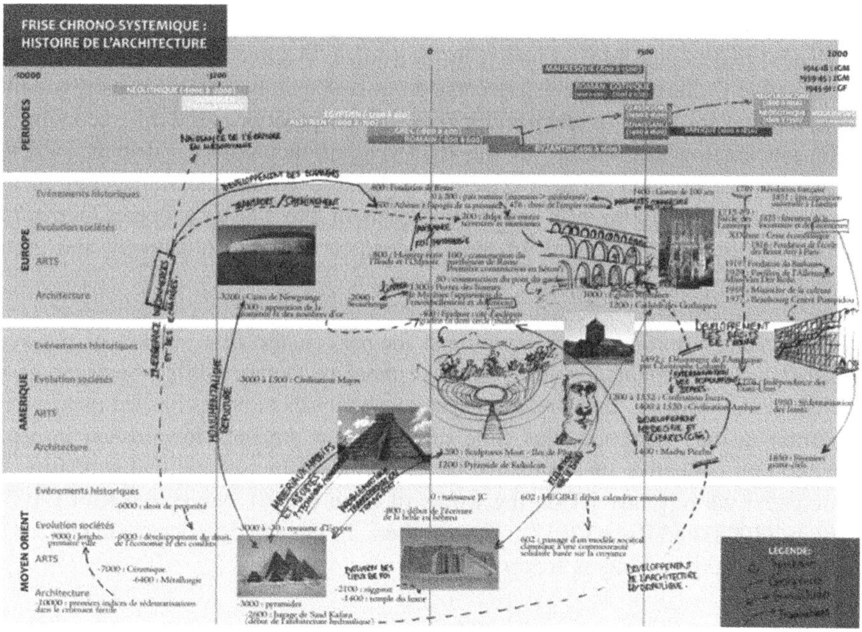

Figure 2.8. *Example of a chrono-systemic timeline (Laura Campelli and Manon Houssin 2019)*

2.4.3. *Master's degree (seminar/studio): working on resilience through mutual aid and the collective*

The abovementioned research work, – particularly when set within the framework of the *Ignis Mutat Res* – had already helped us to develop the teachings of the Master's degree[25], in particular S9 and S10 with the support of IEPs, by placing the students in a position to anticipate 2050 (chosen because of the objectives of various transition scenarios, in particular, the negaWatt and Afterre 2050 scenarios). For the past five years, the work developed in this context has placed strong emphasis on a transcalar approach (through the territorial resource), and an ecosystemic perspective (by an integrative approach towards the necessary changes in practices, in terms of agriculture, food, mobility, health, etc.), while questioning well-being and the pleasure of living. This approach relies on the scientific activity of the research unit and on a very multidisciplinary and interprofessional teaching team[26]

25 Thematic area: "Architecture, environment and constructive cultures". ENSA Grenoble, M2 educational coordinator: Nicolas Dubus.

26 We could even say "meta-disciplinary", since climate change issues transcend disciplinary logic.

which Grenoble's Master's degree (whose curriculum spanning across four thematic semesters) makes possible[27]. Finally, in the course "Architecture, Environment and Constructive Cultures", projects are carried out collectively which makes it possible to experiment and put into practice the values of cooperation, solidarity, economy of scale and sharing know-how. In the first year of the Master's degree, students collectively design and build a concrete realization within the framework of the Design/Build Lab studio[28]; in the second year of the Master's degree, territorial analyses within the framework of the IEPs are carried out collectively and group projects are strongly encouraged[29].

We recently introduced collapsologist theories in the first session of the year with Master's 2 students[30], with works like those of Dupuy (2002), Servigne *et al.* (2018), or Rey (2014). This stream of systemic analyses of crises has led us to a new evolution in our pedagogy, this time orienting our students towards an in-depth reflection on collective mechanisms that can promote mutual aid and solidarity. In this context we rely heavily on the works of Dion (2016), Novel and Riot (2013), Hopkins (2008) and Rivat (2017), among others. In this view, history is mobilized to put ideas and experiments into perspective.

2.5. Conclusion

The various testimonies offered here show that the teaching of history, in relation to climate change, is being mobilized both as a stand-alone discipline and as a privileged dimension of other fields of education. The contexts evoked highlight the singular vocation of a discipline which, in essence, unifies knowledge, whose multiple uses, as we have seen, lend it the versatility of a kind of "Swiss army knife", useful in many ways to support future educational developments.

Its contributions and the changes it triggers by taking environmental issues into account exist at different levels: new approaches (environmental history), new

27 At the ENSAG, six themes (courses) coexist in the Master's degree, supported by six teaching teams for EU studio, seminar and thesis teaching units.

28 Coordination: Marie and Keith Zawistowski. AE&CC, ENSAG.

29 Coordination: Nicolas Dubus, AE&CC, ENSAG. About 10 students work together to analyze the site and highlight the issues, in groups of two or three students for the project design phase. PFE projects on the same site are designed as the building blocks of a global outlook.

30 Theories addressed in the Master's degree, where the seminar enables a direct exchange with the students but we do not address these directly in the Bachelor's degree since the framework of the lecture in a lecture hall does not enable a dialogue with each of the students.

temporalities (landscape) and temporal scales (history of the Earth, the history of Sapiens, history of the climate, etc.), new teaching methods and tools (study of precedents, chrono-systemic timeline, to mention just a few).

It is therefore a significant upheaval, for this discipline as for all those that contribute to the teaching of landscape and architecture, generated by the chaos of today. Knowing the past well in order to tackle the future is a principle that takes on its full meaning at a time when project tools need to be updated, or furthermore, reinvented. Like all the actors of the EnsaÉco network, the initiators of the round table that has been reported here rely on the virtues of sharing experiences and debating ideas in order to face the challenges of the future. Let us hope that exchanges about research and the teaching of architectural history, the city and the landscape – but also the lessons of history in each of our disciplines – will continue to be enriched, enhancing the pedagogy of our institutions and contributing to the urgent changes that we have to make together.

2.6. References

Armitage, D. and Guldi, J. (2014). *The History Manifesto*. Cambridge University Press, Cambridge.

Auzanneau, M. (2015). *Or noir. La grande histoire du pétrole*. La Découverte, Paris.

Baridon, M. (1998). *Les jardins – paysagistes, jardiniers, poètes*. Robert Laffont, Paris.

Bergeret, A., George-Marcelpoil, E., Delannoy, J-J., Piazza-Morel, D. (2015). L'outil-frise : une expérimentation interdisciplinaire : comment représenter des processus de changements en territoires de montagne ? [Online]. Available at: https://hal.archives-ouvertes.fr/hal-01230027.

Berthault, M., Renaudin, A., Vorreux, C. (2019). "Mobiliser l'enseignement supérieur pour le climat" : le nouveau rapport du Shift sur la formation aux enjeux climat-énergie. Le défi : former les étudiants pour décarboner la société. *The Shift Project* [Online]. Available at: https://theshiftproject.org/article/nouveau-rapport-mobiliser-superieur-climat/.

Bonneuil, C. and Fressoz. J.-B. (2016). *L'Événement Anthropocène. La Terre, l'histoire et nous*. Points, Paris.

Choppin, J. and Delon, N. (2014). *Matière grise : matériaux, réemploi, architecture*. Éditions du Pavillon de l'Arsenal, Paris.

Considère, S. and Tutiaux-Guillon, N. (2013). L'éducation au développement durable : entre "éducation à" et disciplines scolaires. *Recherches en didactiques*, 15(1), 111–133.

Coste, A., D'Emilio, L., Guillot, X. (2018). *Ruralités post-carbone. Milieux, échelles et acteurs de la transition énergétique*. PUSE-ERPS, Saint-Etienne.

Dion, C. (2016). *Demain, un nouveau monde en marche*. Actes Sud, Arles.

Dupuy, J-P. (2002). *Pour un catastrophisme éclairé. Quand l'impossible est certain*. Le Seuil, Paris.

Edgerton, D. (2006). *The Shock of the Old: Technology and Global History Since 1900*. Oxford University Press, Oxford.

Findeli, A. and Coste, A. (2007). De la recherche-création à la recherche-projet : un cadre théorique et méthodologique pour la recherche architecturale. Lieux Communs, no. 10, numéro consacré aux "Formes et pratiques de l'activité de recherche", 144.

Fressoz, J-B. (2012). *L'Apocalypse joyeuse. Une histoire du risque technologique*. Le Seuil, Paris.

Gras, A. (2007). *Le choix du feu : aux origines de la crise climatique*. Fayard, Paris.

Grimal, P. (1964). *L'art des jardins*. PUF, Paris.

Harari, Y.N. (2015). *Sapiens: A Brief History of Humankind*. Vintage, New York.

Hopkins, R. (2008). *The Transition Handbook: From Oil Dependency to Local Resilience*. Green Books, Cambridge.

Hopkins, R. (2019). *From What Is to What If: Unleashing the Power of Imagination to Create the Future We Want*. Chelsea Green Publishing, White River Junction, VT.

Illich, I. (1974). *Energy and Equity*. Harper and Row, New York.

Kottek, M., Grieser, J., Beck, C., Rudolf, B., Rubel, F. (2006). World map of the Köppen-Geiger climate classification updated. *Meteorol.* Z., 15, 259–263.

Krautheim, M., Pasel, R., Pfeiffer, S., Schultz-Granberg, J. (2014). *City and Wind: Climate as an Architectural Instrument*. DOM publishers, Berlin.

Lamarck, J.-B. (1830). Système analytique des connaissances positives de l'homme, restreintes à celles qui proviennent directement ou indirectement de l'observation [Online]. Available at: https://gallica.bnf.fr/ark:/12148/bpt6k86264v/f157.image.

Massard-Guilbaud, G. (2010). Bilan historiographique de l'histoire environnementale. Speech given at the Assemblée générale de l'Association des historiens contemporanéistes de l'enseignement supérieur et de la recherche (AHCESR), November 6 [Online]. Available at: https://ahcesr.hypotheses.org/.

Massard-Guilbaud, G. and Mathis, C-F. (2019). *Sous le soleil : systèmes et transitions énergétiques du Moyen Âge à nos jours*. Éditions de la Sorbonne, Paris.

Mathis, C.-F. and Pépy, E.-A. (2017). *La ville végétale : une histoire de la nature en milieu urbain (France XVIIème–XXIème siècle)*. Champ Vallon, Ceyzérieu.

McNeill, J.R. (2001). *Something New Under the Sun: An Environmental History of the Twentieth-Century World*. Penguin, London.

Ménard, R. (2014). "Le pic de l'architecture". In *Matière grise : matériaux, réemploi, architecture*, Choppin, J. and Delon, N. (eds). Éditions du Pavillon de l'Arsenal, Paris.

Ménard, R. (2018). "Énergie, Matière, Architecture" – Architecture in VAE (Validation of the Acquisition of Experience, prior learning recognition). PhD thesis, École d'architecture de la ville & des territoires, Paris.

Morris, W. (1891). *News From Nowhere: Or, An Epoch of Rest, Being Some Chapters from a Utopian Romance*. Reeves and Turner, London.

Nash, R. (1967). *Wilderness and the American Mind*. Yale University Press, New Haven, CT.

Novel, A.-S. and Riot, S. (2013). *Vive la co-révolution ! Pour une société collaborative*. Gallimard, Paris.

Quenet, G. (2014). *Qu'est-ce que l'histoire environnementale ?* Champ Vallon, Seyssel.

Revault d'Allonnes, M. (2012). *La crise sans fin : essai sur l'expérience moderne du temps*. Le Seuil, Paris.

Rey, O. (2014). *Une question de taille*. Stock, Paris.

Rivat, M. (2017). *Ces maires qui changent tout, le génie créatif des communes*. Actes Sud, Arles.

Servigne, P., Stevens, R., Chapelle, G. (2018). *Une autre fin du monde est possible. Vivre l'effondrement (et pas seulement y survivre)*. Le Seuil, Paris.

Stein, R. (1977). *Architecture and Energy*. Anchor Press, New York.

Vernet, N. and Coste, A. (2017). Garden cities of the 21st century: A sustainable path to suburban reform. *Urban Planning*, 2(4), 45–60.

3

Ecological Transition Architecture Week (ETAW): An Introduction to the Assessment of Risks and Opportunities

3.1. Introduction

Since 2013, Ecological Transition Architecture Week (ETAW) at the *Ecole Nationale Supérieure d'Architecture de Nancy* (ENSA Nancy) has been a teaching event entirely devoted to the contemporary issues of ecological transition. This week-long event engages the efforts of all second year students. The goal is to make students aware of the challenges of ecological transition in the Anthropocene era over a short and intensive period, and to understand possible levers of action by working on concrete situations at different territorial and temporal scales.

Various exercises run in parallel during the event: the daily press review; analysis of a territory sample as a team; debates organized after the collective viewing of short documentaries; conferences. The students come to understand architecture, and more generally, the inhabited and imagined places, through their temporalities, functioning and interactions, for the sustainability of the contemporary inhabited world and its ecological transition.

The ETAW aims to encourage students to adopt a systemic approach, promoted by the analysis of territory samples (approximately 25 samples per year), in small groups. The proposed territory samples are generic elements, which can be found in any conurbation, but in different forms, which, put together, facilitate an intelligible

---

Chapter written by Alain GUEZ and Antoine STECK.

For a color version of all the figures in this chapter, see www.iste.co.uk/aithaddou/transition.zip.

reading of the territorial metabolism. The student groups propose a synthetic evaluation of the elements in favor of or hindering the ecological transition of the studied situation. A collective reporting activity at the end of the week enables students to understand the situation they have studied as part of a larger territorial system, and better apprehend its different constitutive parts.

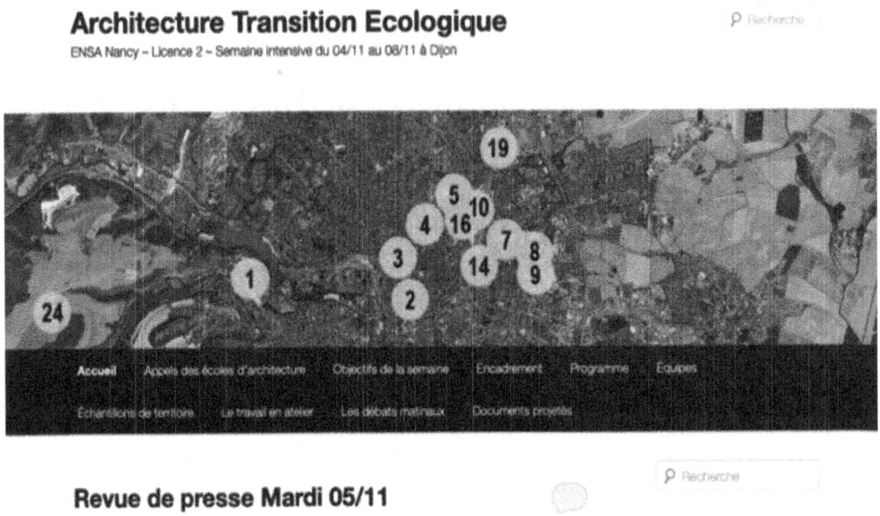

Figure 3.1. *Screenshot of the ETAW Blog home page with the location of the territory samples studied by undergraduate students from ENSA Nancy in the Dijon metropolitan area, as part of ETAW 2019–2020*

It also aims to ensure that the students will understand the importance of reasoning at different territorial scales, starting from concrete situations, and according to different horizons and multiple temporal scales.

The teaching engages theoretical reflections on contemporary ecological issues, in the Anthropocene[1] era, and encourages students to identify concrete translations in the territory, city and architecture. Through a progressive and plural approach,

1 The neologism "Anthropocene" amounts to considering that we have entered an era in which the effects of human action are transforming the planet to an extent that is becoming geologically significant. Whatever date it is possible to trace it back to, this new era would manifest itself today in rapid transformations of the "Earth system", on an unprecedented scale, and which could potentially become irreversible at a human scale. Excerpt from "Is mesology another paradigm for the Anthropocene?" International conference that took place in Cerisy-la-Salle, August 30–September 6, 2017.

accompanied by professionals in spatial planning and architecture, students are required to draw up an inventory of concrete situations, in terms of opportunities and risks, taking into account the challenges of ecological transition.

This week is aimed at all bachelor's degree second year students. ENSA Nancy has chosen to introduce compulsory teaching for a dedicated week in the middle of the bachelor's degree for all students. This positioning in training and its compulsory nature for all students makes it a major educational moment, fully integrated into the first cycle's initial training. This dedicated week is a privileged moment which finds echoes in other theoretical and practical teachings in the educational program of the school (at the bachelor or master degree's levels), in Hmonp or for doctoral studies.

As part of this dedicated week, various exercises are carried out in parallel: territorial actors conferences on the challenges and policies of sustainable development and ecological transition strategies; a daily press review; debates organized after the collective viewing of short documentaries around three successive thematic entries: cycle, energy, living; progressive analysis (in groups of four to five students) of a territory sample to assess and debate the risks and opportunities which can be identified in the situations studied.

The purpose of these exercises is to support students in progressive work, encourage them to work as a team, enable them to understand the complexity of the challenges of ecological transition, in particular, through a thematic approach that encourages them to reason at different spatial and temporal scales, about both local and global issues.

3.2. Identifying current topics in terms of ecological transition

The students were divided into 25 groups of four to five students and invited to produce individual and collective work. Every morning, the school provides a choice of newspapers: national and local dailies, for a press review team made up of one member from each of the study groups.

A teacher organizes the press review work, which is structured as follows: distribution of national and local newspapers, individual reading of part of the press, identification of themes relating to ecological transition issues, collective debate on the topics identified, collective selection of subjects, assembly of a significant press review about the questions raised by the media.

Figure 3.2. *Choice of presentation panels for daily press reviews, produced by 2nd year bachelor degree students at ENSA Nancy – ETAW 2019–2020*

This exercise takes place throughout a quarter of a day and enables students to identify issues, as well as subjects relating to ecological transition. These may be economic, geopolitical, scientific, or even social topics, etc., which, at first glance, may seem far from architectural, urban planning or landscape issues, but whose links with their discipline appear clearly to them in this exercise.

Although the non-specialized press deals with subjects which are not strictly architectural, urban or even landscape-related most of the time, it invites students, as part of this exercise, to make an effort to translate the identified subjects into their field of competence and more particularly, in relation to the building environment that structures contemporary lifestyles.

Depending on the news, ETAW could be set alongside national or international events, as in 2015, during the Cop 21. This concomitance made it possible to provide rich material for reflection, relayed by the press, and to become aware of the different scales and themes of reflection jointly necessary to address the challenges of ecological transition.

3.3. Territorial system sampling

The territory samples proposed are generic elements, which can be found in any agglomeration, but in different forms, and which, when put together, facilitate an intelligible reading of the territorial system, and more generally, of contemporary ecological systems.

In order to enable students to understand the risks and opportunities in terms of ecological transition, territory samples (25 samples in all) are first identified by teachers. Each sample is given to a group of four to five students.

The territory samples chosen are grouped into four large families of generic "objects": housing (collective, individual, of different types, built at different times and located in different urban, peri-urban, rural contexts); large facilities (universities, hospital centers, stadiums, etc.); specialized areas (commercial, activity, food production, etc.); transformation centers (industry, sand pit, recycling center, cement plant, etc.). Throughout the week, each group of four to five students works on one of these samples, successively analyzed in terms of cycles, energy and the living.

It should be noted that among the samples chosen, some may have remarkable ecological qualities. For example, the garden of the Lortie association, located on the urban edge of the metropolis of Nancy, member of the Cocagne Network[2], promotes organic agricultural production with short-circuit distribution and employs people undergoing professional integration processes. This virtuous sample of the territorial ecological mesh raises the question of agricultural production and food distribution to a greater extent, beyond the experience of this unique case.

At another level, among the samples collected, the treatment plant raises the question of water cycles, their treatment from upstream to downstream, the management of rainwater, the consumption of drinking and non-potable water, the management of urban effluents, etc.[3]. Or still, a shopping mall poses the question of the underlying automotive system, dependence on oil, soil waterproofing, the relationship between urbanity and commerce, or the future of urban centers, etc. For the students, it is a question of deconstructing the sectors, practices, planning and infrastructures on which they depend and seeking alternative, resource-efficient, sustainable (permanent/reversible/adaptable) models to better criticize the samples and eco-systems studied, from the viewpoint of ecological transition.

Thus, the territory samples offered to students raise many questions that they discover through the surveys they conduct throughout the week. These samples serve as a starting point for in-depth field studies and raising discussion topics.

---

2 "Inspired by a Swiss model to which a social component was added. In 1991, the first Jardin de Cocagne was born in Chalezeule (25), supported by the Julienne Javel association. Faced with the exclusion which then affected people of rural and agricultural culture, this new concept, initiated by Jean-Guy Henckel, diversified integration activities through economic boost and extended the offer to bring people experiencing difficulties back to work." Excerpt from the Réseau Cocagne website. Available at: http://www.reseaucocagne.asso.fr/origine [Accessed on August 21, 2020].

3 On this subject, see Courrier International, Water Atlas, Special Issue, September–October 2020.

During the final report, the work carried out on all of the samples is presented, per thematic family, to the entire group of students. The students can thus reconstitute a more complete territorial system, in which the sample they have studied fits in more specifically.

3.4. From a thematic analysis...

In order to guide the students in their assessment of the risks and opportunities of their territory sample in terms of ecological transition, a thematic approach is structured to promote a progressive construction of their analysis. The choice of the three thematic inputs – cycles, energy, the living – aims to structure the understanding of ecological issues. These three themes certainly overlap, and they are systemically related to one another. Whether in social, economic or environmental terms, they prompt a dynamic understanding of the territory and the architecture.

The groups of students are asked to successively analyze their territory sample through the prism of a theme. Each theme is developed by a team throughout a day, who must then coordinate their interpretations in order to produce brief analysis documents for their territory sample (one horizontal A4 page per student, divided into one half devoted to iconography, and the other to the development of the argument). Students are not only asked to analyze their sample depending on themes, but also in terms of periodization, establishing differences between yesterday, today and tomorrow. On the one hand, this temporization enables students to understand the dynamics of territorial development, spatial planning and the ecological impact of the techniques used, and also to identify, especially online, cases making it possible to put the samples they are studying into perspective, by comparing them with virtuous examples which take the challenges of ecological transition into consideration.

Working in groups of four to five students is necessary in order to deal with the daily theme submitted to them and distribute the work so that their individual productions are complementary with one another. This involves both individual and team work, aiming to promote exchanges and debates between groups of students and with their teachers.

A day of production is devoted to each of the themes. In order to introduce the thematic approach, a half-day of debate is organized on the basis of short documentaries, so that the students can understand the challenges of each of the themes[4]. The documents presented to the students are either extracts from

4 For example, on the question of cycles, five short documents were presented and used as a support for debate with the students: *Introduction à l'économie circulaire*, by François

documentaries[5], or Mooc extracted from the Uved[6], which have the advantage of presenting analyses at both a global and a local scale, theoretical and concrete in nature. The projection of these documents prompts collective discussions led by teachers promoting ownership and debate within the class.

Following this collective session, the discussions are continued in smaller groups of around 30 students, facilitating the exchange on the subjects and arguments developed in the documentaries and interviews projected.

Thus, rather than offering a rigid framework of analysis, the press review and the documents projected and discussed enable students to sharpen their gaze and their critical sense, as well as to build an analysis of their territory sample based not only on what is observable in the field (identification, interview with actors, study of documentation), but also on what is relayed in documents, available online for the most part.

During this dedicated week, the purpose is to learn to question oneself, document oneself, navigate through and discriminate from the available information, in other words, learn to learn and find a solid point of reference in a large body of information.

3.4.1. *... towards the assessment of risks and opportunities in terms of ecological transition*

In order to enable students to pool their knowledge and learn from the survey collection performed during the week, a blog is built[7], bringing together all of the information related to the week's unfolding: reception, call from schools of architecture[8], educational goals, supervision, program, teams, territory samples, workshop work, morning debates, projected documents. The blog is also the archiving medium for individual and collective work (press review, thematic

Grosse, 10'; *La consommation collaborative*, by Anne-Sophie Novel, 13'; *Villes soutenables*, by Sabine Barles, 8'; *Transitions urbaines à la durabilité*, by François Mancebo, 8'; *Demain*, Chapter 3: *l'économie*, 28'.

5 The film *Demain* (2015) by Cyril Dion and Mélanie Laurent was notably used and presented by the chair.

6 Université Virtuelle Environnement et Développement Durable. Available at: https://www.uved.fr/.

7 See: http://blogs.nancy.archi.fr/transition/.

8 "Appel de Nancy" : Enseigner l'architecture et l'urbanisme à l'ère du développement durable. Available at: http://ensaeco.archi.fr/. For teaching ecological transition in architecture and landscape schools using collaborative and benevolent teaching practices.

analyses, final report). This support medium makes it possible to organize a global report in half a day: the students post their production there, and it is then directly accessible online for the final collective report.

The final report of the students' work is a short, five-minute presentation per group, where each group presents its assessment of the risks and opportunities in terms of the ecological transition of their territory sample, illustrated by two slides: one for the risks and the other for opportunities.

This assessment is developed by each group during a final special day, fed by thematic analyses (cycles, energy, the living) developed throughout the week. The presentations follow, making up four groups corresponding to the families of chosen samples (housing, large equipment, specialized areas, processing devices).

Figure 3.3. *Final presentation by a group of undergraduate students from ENSA Nancy in front of all of the witnesses – ETAW 2019–2020*

This structuring of the final presentation enables Major witnesses to discuss the evaluations developed by the students and also all of the bachelor's degree second year students to learn about the work of their colleagues in half a day, and to situate their own sample in a larger territorial system.

Throughout these exercises, students are invited to discover notions, concepts, principles, forms, poetics, techniques, processes, approaches, attitudes, etc. for the sustainability of the contemporary inhabited world and its ecological transition. Students are encouraged to understand architecture and more generally, inhabited and imagined places, through their temporalities: durations, rhythms, cycles, permanence, operation, production/consumption, resource renewal, different uses.

3.5. A framework for reflection prior to the project

Although at ENSA Nancy, as in other French schools, the core of the pedagogy is oriented towards the practice of the project, ETAW offers a reflection and tools upstream of the project's conception. The assessment, in terms of the ecological transition of the territory samples progressively constructed by the students, forms an assisting framework for the design, which can later be summoned for the project itself.

Thus, in a short but intense and dedicated time, students are required to learn to analyze a situation with a multi-scalar and multi-thematic positioning, in order to shift towards an architectural and urban diagnosis, a necessary prerequisite for any habitat transformation project. In particular, one of the ambitions of this week is to begin to make students aware of the notion of interlocking scales, the understanding of which is necessary for an ecological approach.

ETAW is an opportunity for the individual and collective construction of a shared observation of the city and of contemporary territories. It is also an opportunity to compare this observation with professionals and elected representatives, who are currently working for the development of territories through four meeting formats: conferences, debates, workshops and the final report. This enables students to begin to discover the ecosystem of territorial development actors, which they will become a part of at the end of their training. For this, a strong partnership with a metropolis[9] and interventions by architects, technicians and elected representatives enables ETAW to anchor the questions it raises in reality.

9 Since the first year of the implementation of this pedagogy, ETAW has been able to develop in the metropolis of Nancy, but also, in conjunction with engineering schools, both at Metz Métropole, and in 2019 at Dijon Métropole, within the framework of "agreements with local authorities, providing for the onsite reception of students, in particular".

4

Climate Commons – Educational Research in Action

4.1. Introduction

This chapter is based on a project in which research is taking tentative steps towards education and can therefore project itself into action: project work with "profound goals" in "real time" about two overlapping topics which appear inseparable: the climate and the common.

Here we are in the third year of the bachelor's degree at the Paris Val de Seine School of Architecture, an interdisciplinary team on a multi-year work cycle combining project teaching, bioclimatic training, and digital modeling on programmatic issues which appear to us, in their openness and by their very indeterminacy, as "ideal" opportunities to reflect upon the climate and the common. In the semantic and experiential shift ranging from teaching to learning (that is, learning to learn), a considerable game is formed, in which the difficulty of knowing the arrangements for organizing the community of "learning/doing" beforehand reveals the need to consider the climatic metabolisms of the architectures which house them.

We are engaged in a collaborative sequence[1] with *Le Fresnoy* (*Studio national des arts contemporains*, the National Studio for Contemporary Arts in Tourcing,

---

Chapter written by Edith AKIKI and Emmanuel DOUTRIAUX.

For a color version of all the figures in this chapter, see www.iste.co.uk/aithaddou/transition.zip.

1 This has been a three-year-long collaboration as the partnership was reconducted in 2020–21, and will be followed by an exhibition of the overall didactic and research process in the autumn of 2021 at *Le Fresnoy*, as part of Panorama 23.

France – Bernard Tschumi Architects) which reconsiders the future of the building and the educational/production institution in close relation with its teams. This, under two principles, we have called "*comforting*" (updating the ambiance as well as the living conditions) – where once again the climate and the common are combined – and "*upgrade*" (of the institutional program now at the crossroads between the sciences and the visual arts) – where an opportunity to rethink the methods of architectural design itself arises.

At the interface between project work and dynamic bioclimatic simulations, the opportunity arises to reconsider architectural *poiesis* – by what is the project governed? What tools should be used? How can we think about the art of presentation (Daston and Gallison 2012) to which we are invited today? What are the politics of common societies with which it is intertwined? What are the affordances? Which project? What governance? (Coriat 2015; Dardot and Laval 2014)). In this knot of favorable constraints, where the "making space/leaving place" of architecture (Hertzberger 2011) is at play again, there is a formidable didactic stake: how communicate to our "audiences" what means architecture now has to make do with.

> We generally think of buildings as very particular designs. I envision them as extremely porous systems, exchange systems, air flow systems, systems of air passing from the inside to the outside, facilities where one can be sheltered from the wind or encounter the wind, feel the air, the seasons, the light, the sun. It is not fullness that must be built, but a vacuum. It is not the bricks that matter, but the spaces, the air, the atmospheres, the climate, the inhabitants and their movement. (Vassal 2012)

4.2. Climate

The paradigm shift in ecological issues, at a time when the richness of living things in the biosphere is collapsing, has a profound effect on our conception of the world and our economy (*οικονομία*) – the administration and shared management of the resources of our common home, whether we consider it at a planetary or an architectural scale.

The environmental issue here brings together our respective expertise, within which the climate argument is key. We work on the project in two ways:

instrumented (addressing the measurable) and *attentive* (in accordance with a sensitive approach)[2].

In the most objective/objectified sense, the work examines the pressure exerted by the environment on architecture's metabolism, and conversely, the way in which the latter affects the environment it is part of.

This is a matter of data on exogenous seasonal and daily cycles (wind, sun, rain) as well as endogenous cycles (inhabitants, materials, equipment). This sets out the project as the simulation of possible/plausible behaviors of this living being made up of meteorological, architectural and human flows. This involves putting an end to the conception of design as the projection of a stabilized form – of an essentially inert state – and, on the contrary, to consider it as the *duration* (in Bergson's sense) of a continuous performance, that is to say, during and after the moment of delivery.

The improbable but necessary objectification of these actions is coupled with another uncertainty related to the fact that a constructed situation should comfort us or discomfort us, welcome us or indispose us, and to the discreet attention given to the subtle mechanisms of sensation and perception. This surpasses the sensory level, figuratively hitting a nerve in this more or less well-established climate. What we usually call its *atmosphere* or *ambiance*.

How is the design time able to integrate the complexity of sensitive issues which are difficult to predict, since they arise from various subjectivities? Should the project not be considered within a horizon of uncertainties, as the incubator of possible sensations, the facilitator of future situations, the activator of atmospheres it cannot fully master?

This is difficult to conceive of, but invites a form of humility, and asks us to think about the project as a device offering a myriad of possible climates and atmospheres, which can be activated by users on demand.

---

2 Here, the fertile balance is exposed between a physical approach to climatic phenomena, or environmental phenomena in a more general way, and their phenomenological approach – this balance being notoriously expressed in the research work of the Cresson team at Ensa Grenoble. But the logic of attention also refers to the work of Yves Citton (2014).

4.3. The common

The common is not innate, but acquired. It is the subject of a project. And it represents, for and by a given community, the co-management of its resources, the adherence to its related rights and obligations, eliciting shared governance. This second line of work – no less affected than the previous one by the paradigmatic breakthrough of the Anthropocene – implies the inclusion of architectural design in new democracies of use.

Without delving into the scientific controversies related to the "return of the common(s)" (Coriat 2015; Dardot and Laval 2014) – for this is not the central object of our discussion – let us remember that this concept can cover very different meanings, depending on the legal foundations on which it is based[3], the degrees of reification assigned to it[4], and depending on whether it relates to its original land use or to its increasing contemporary informational dimension[5].

Let us consider two reasons for the importance of this return of the common to project design. This question brings into play, displaces and goes beyond the controversial reality of the public space (of which the user appropriation often highlights the flaws of the authority which governed its design), moving beyond the conventional lines dividing public and private domains. Under the cover of architectures of *(un)certain quality* and of *a certain size* ("rather large"), this approach makes it possible to renew, to a relative extent, and to reinvent, to a larger extent, the conditions of existence of ancestral commons – such as those historically offered outdoors – by means of the (environmental, climatic, aeraulic) liberalities which are dispensed therein.

However, the profiling of these training, teaching and *learning* places where new behaviors, new postures and new relationships are being invented every day before the activities themselves, appears more difficult than ever. The programming of

---

3 Zask (2011) has clearly drawn a difference between a liberal Anglo–American culture (where the common feeds almost naturally from a "communitarian" conception of society) and a German–Latin culture (where the common is inscribed against the idea of an organic community of which the Nation would be the depositary).

4 Dardot and Laval (2014) vigorously disagree about a reification of the common, of which they consider certain supporters of the ecological cause to be victims, the latter confusing natural commons with the common, which would relate precisely to an artifactual dimension, and to a political principle governing production.

5 Coriat (2015) analyzed the way in which the "return of the commons" updates the economist approach offered by Elinor Ostrom, by showing how a mechanism is reversed by moving from land (the limit guaranteeing protection) to the digital (openness protects information by multiplying it).

these places, relatively unconstrained from a technical viewpoint (nothing to do with the complexity of hospitals or performance halls, and their strict functional prescription, for instance), is more uncertain than ever, symptomatic of an uneasy prefiguration of uses. This uncertainty of the programs, and the relative indefiniteness of the resulting achievements, both extensive and indecisive, echoes the availability of ancestral commons, their poorly established proprietary logic, and their varied uses.

These new "rolled-up indoors" commons (whatever they may be: learning centers or platforms, *learning commons*, *learning centers*) can also be understood as a new way of conceiving the socializing spaces for which the structure of historic campuses has for half a century been insufficient.

But in a specific case like that of *Le Fresnoy*, this common approach could finally be the opportunity to re-explore the capacity of a public institution to open up more widely to its immediate environment or to embody a form of porosity which has not been the trademark of the original project (far from it). Moreover, it is a relatively "weak" meaning of the concept, little-asserted conceptually, that we adhere to in this educational adventure. Not emerging from "formal" or declarative commons – inhabited by models of the social and solidarity economy, or informed *open source* ecology (participatory housing, production cooperatives, *creative commons*) – nor "eruptive commons" – urban struggles as initiated by the *Indignados* or the *Zadist*[6] movements – we are interested in the impure, informal, discreet commons. They are "diffuse" and made possible, or *afforded* (Gibson 1979) through "open" programs, equipment and networks. They pool a resource and generate relationships without a charter of uses necessarily being in place, and without really formalizing the sharing of its governance.

4.3.1. *Climate and the common*

The common and the climate. The climate and the common. The specificity of our proposal lies at the crossroads of these issues, where technical, aesthetic and political arguments are tied.

Unlike the modern "plastic" approach of the project and its (belief in) the mastery of a stable and isotropic space, in its fixed projection, the function of the project here relies on plausible scenarios in relation to general objectives. It models and simulates its implementation regarding a range of means deployed to do so. The

6 "Zadist" is a French neologism, from "Zad" ("*Zone à défendre*", which means "area to defend"), which is a subverted version for the planning lexical/ideological form of "*Zone d'aménagement différé*", or "deferred development zone".

spectrum of possibilities remains open, because the characteristic of the atmosphere, or ambiance, is its variability, and the "regulability" of its control.

But this project-based procedure also calls for an examination of its results with regard to the general objectives that it initially stated. That is, its assessment, or its valuation, in Dewey's[7] sense. Because the purpose of this valuative protocol, in support of the project's new *instrumentarium*, was to reach not only to an extended college of experts, but also a wider circle of users (or resident experts); in other words, it brought forward the relationship between "how to plan" and "who decides".

Thus, the same projectual dossier would include external mesological solicitations (climatic agents, urban environment, etc.), the behavioral response of the building itself (depending on its levels of protection, its regulatory capabilities, etc.) as well as the anticipation of its inhabitants' actions (in relation to their modes of action, user scenarios, events they initiate). In this way, the common and climate are closely related.

Figure 4.1. *The famous* Fresnoy *(source: Bernard Tschumi Architects, photographed by Peter Mauss)*

7 The notion of *valuation* is central to Dewey. It makes it possible to articulate the perception resource, where emotions and desires are appreciated, where experience is played out, and the evaluation resource, where ends are valued in relation to initial intentions. What he calls "survey" will be the public dimension of this evaluation (Dewey 1927).

Le Fresnoy is famous for the heroism of its architectural posture: the enhancement of unique heritage (this former palace of workers' entertainment), the scenography of a heroic rooftop (giving rise to the imagination of an *in-between*[8] two worlds). It is therefore an all the more imposing challenge to model or simulate a possible third state of the built environment, which includes it in the challenges of the climate and the common.

Is it not a question of creating new sensitive expressions – exceeding the mere register of the visible – which are in phase with this architectural generation whose instrumentation is no longer projective, but involves simulation, whose place is not the space, but the ambiance?

Working with/on the *Studio national des arts contemporains du Fresnoy*, the question of the relationship between architectural (or artistic) design and its digital *instrumentarium* came to the forefront. Working at *Le Fresnoy* is not completely neutral for us, since the quality and the reputation of its production, in the register of new digital visualizations in particular, challenge the artistic world as a whole, and also seem to question the capacity of the scientific world to make images and simulations free from the representation regime in these times of "exerted judgment" (Daston and Galison 2012).. This is what the exhibition *Le Rêve des formes* (The Dream of Shapes), celebrating twenty years of the *Fresnoy* institution, was already staged at the *Palais de Tokyo* in Paris in 2017.

Inscribed in the production, in the exercise of learning in *Fresnoy*'s creative setting, a transitory and transactional nature of the image is worked upon, both in its speculative and its prospective dimension. This paves the way for new hardware interfaces. In return, this experience calls into question our way of thinking about architecture, our way of considering design, of producing the project, of producing images about it, which in turn makes us reflect about our training practices.

From then on, it is no longer two, but three threads which intertwine and run in this way in the course of our approach. They relate to the climate, the common, and the very nature of design.

8 The figure of the *in-between*, while not completely new historically speaking, nonetheless plays a major role in the architect's doxa, between what it potentially accommodates in a series of buildings to be construed, and an educational nature, zones of programmatic uncertainty, even non-programs, slipped between architectural layers of more or less protective qualities. One of the challenges proposed to our students will consist of re-examining this issue of the in-between with regard to contemporary bioclimatic expectations.

4.4. Educational sequences

"The international *Fresnoy-StudioLab* will be a higher education and research establishment, with the mission of activating fruitful exchanges between the arts, sciences and social reality, and in collaboration with close institutions, of becoming a 'regional network lead', with a national and international vocational influence.

While the relationship between the arts and sciences is generally conceived by its promoters from the point of view of what scientific and technological knowledge can bring to artistic creation, it appears to us that, reciprocally, scientists may benefit from the vision and the imagination of artists, as well as from dialogue with society. Where scientist refrains from extrapolating research and discoveries beyond the specific discipline's field, the artist can offer transpositions, extensions, an opening to aesthetic and social perspectives, which in turn can stimulate purely scientific research[9]."

Figure 4.2. *The famous* Fresnoy *(source: Bernard Tschumi Architects photographed by Peter Mauss)*

9 Alain Fleischer, Director of *Fresnoy*: "Le *Fresnoy-StudioLab* international" project, Fall 2018.

In reality, with a fairly modest increase in the surface area of 1,500 m^2 compared to the 10,000 m^2 currently available, the request essentially seemed to concern an issue of visibility and symbolic representation (to highlight the relationship between the digital arts and research and engineering sciences, under the *increase* mentioned by the director of *Le Fresnoy*), as well as an issue relating to amenity and comfort, which amends the original industrial referent, to also make the institution a real living place (learning through relationships, valuing the common).

This endeavor had to do with a reality historically unfolding into two consecutive ages (*Le Fresnoy 1*, fanciful palace of entertainment; *Le Fresnoy 2*, heroic studio of contemporary arts), which was complex to deal with in its prosaic reality. What the daring party of the Tschumi project authorized with one hand through the radical nature of its posture (protection of the market halls as they were, strong identification of the *École*), it did not make very easy in everyday life (the climatic rigors of the in-between, the fragility of a highly exposed technical device, issues of the lighting of programs, relative distributive complexity). The diagnosis that we were able to collectively establish was that of a critical situation, due to the highly costly energy/climate design of the original project, completely out of touch with the societal context and current norms.

How could the radicalism of an architectural figure be matched with the requirements of a fluid distribution of the building/equipment resources? How would the equipment also be able to identify a community at work – albeit fleetingly (the time of a few semesters spent together)? How to make soft forms of appropriation possible? In terms of the common issue, this projectual endeavor certainly arose.

How can we endow this building project with a qualified relationship with meteorological stress? How can we interpret the large coverage potentials, the outsourced networks, and the "in-between" (Gibson 1979)? How can we question the atmospheres resulting from this attention? From the point of view of climate, it was a real challenge.

For this reason, the climatic approach was implemented in three stages. First, a diagnostic sequence, in the form of a bioclimatic inventory integrating seasonal thermal amplitudes, wind directions and the course of the sun depending on the seasons, the casting of shadows. The recommended software, in open access, was the following, for modeling the wind, sunshine and solar energy respectively: Flow Design, SketchUp, ArchiWIZARD.

This first approach to the existing situation was accompanied by a qualitative analysis of the condition of the facades and the distribution of programs and flows in plan. We integrated the initial responsive hypotheses to these challenges throughout the work on the project. In addition to this, a sensitive analysis of the site had to be carried out, in terms of comfort (heat/cold, light/dark, weight/lightness, closing/opening, smoothness/roughness, hardness/softness, etc.), this approach being facilitated in season 2 (2019–20) by a preliminary intensive on-site immersion week (and initiation to image production tools by local technical teams).

In the development phase, with regard to the specific climatic challenges of each project and the hypotheses formulated, interfaces could be operated on the levels of lighting, the resources and energy networks, but without validation, only operate by calculation, given the complexity of the situation, and the limited duration of the exercise (one semester). This work could have been carried out by collecting information from 2D (Autocad) documents from often heterogeneous generations at the cost of full digitization (under Bim Revit) of an extraordinarily heterogeneous site, with volumetric aggregations and often very complicated paths to be untied.

With the help of the digital model, the concluding sequence gave rise to the simulation of aeraulic behavior and sunshine, an energy/ventilation approach, and to the production of immersive views in the project (video and VR) making it possible to imagine the transformation of lighting levels (in pre-existing volumes) or their *ex nihilo* qualification with a certain degree of accuracy. At the same time, it would also take into account the prevailing relationship for qualifying the possible atmospheres and "modes of existence". Thus, in various ways, this in-between showcasing the conceptual mark of Tschumi's building (but which has also proved to be its energetic Achilles heel) was questioned both in terms of its bioclimatic potential, and of its capacity to be put to real as well as atypical uses.

"The Factory" – Tony Moreira da Silva proposes a densification of the in-between, which takes charge of the new programmatic deal in fragments. Parasitizing/colonizing the existing entities with a series of occasional grafts, offering their work or relaxation amenities to the otherwise little-transformed production/distribution surfaces located at their foot.

This is a series of "factories" revisiting a new industrial vernacular generation (natural ventilation in a soft bioclimatic mode operated by extraction turrets), slipped on the valleys of the halls and under the shelter of the Tschumi roof.

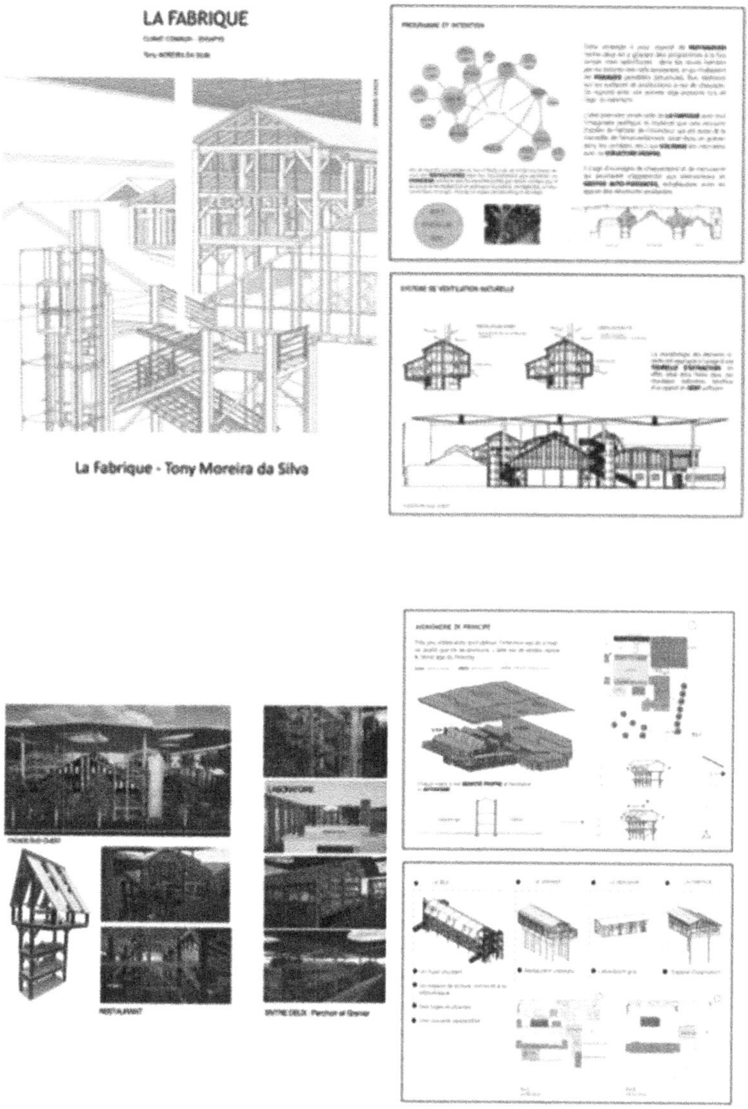

Figures 4.3 and 4.4. *Project by Tony Moreira da Silva*

"The Nave: Enlightening the Public" – Margaux Lefevre envisions a new distributive line, transgressing the nature and function of the current *Fresnoy*. By redistributing the programs, it is about providing clarity, both literally and figuratively. Thus, the stroke of the blade given to the axis of the construction turned the dark circulation into a luminous nave, giving access, on its north side, to a hive of exchanges between users sheltered by large luminous sheds. This distributive strategy was also conceived to make the central axis available to the public, by transferring restricted access to services to its sides.

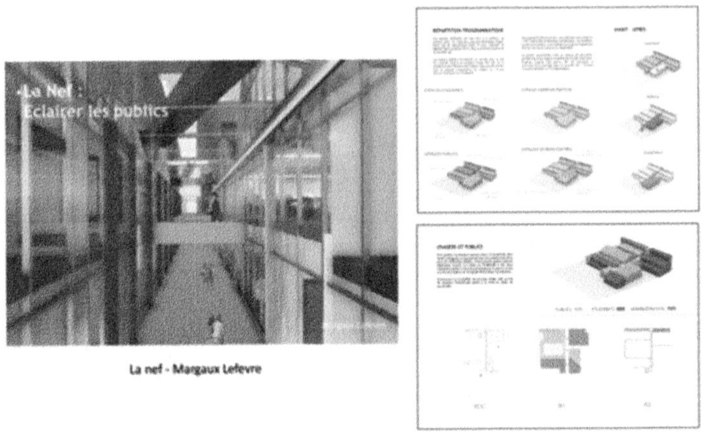

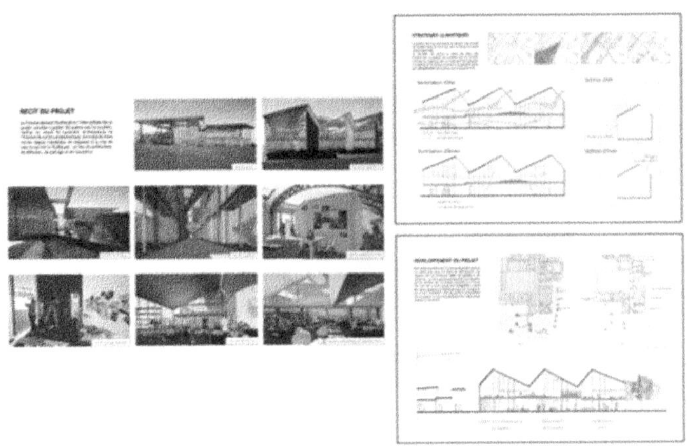

Figures 4.5 and 4.6. *Project by Margaux Lefevre*

"**On top/on bottom**" This is the qualification given by Pauline Brochet and Luke Gates to a project functioning as a soft subversion of Tschumi's enterprise. This was achieved by subverting the energy networks previously exposed in the in-between and turning them into technical galleries as an underpinning. This was achieved by drilling the ground of the square to conduct the energy of a Canadian well, exhuming on this occasion the historical layer of the old swimming pool for which they reserved various scenario configurations. A new bandwidth was invented and reserved for production (including StudioLab) on the SW side of the equipment, served by a distribution functioning as a passive sensor and looping at its top the network of gateways from an in-between, lightened in its technique and strengthened in its ambulatory vocation.

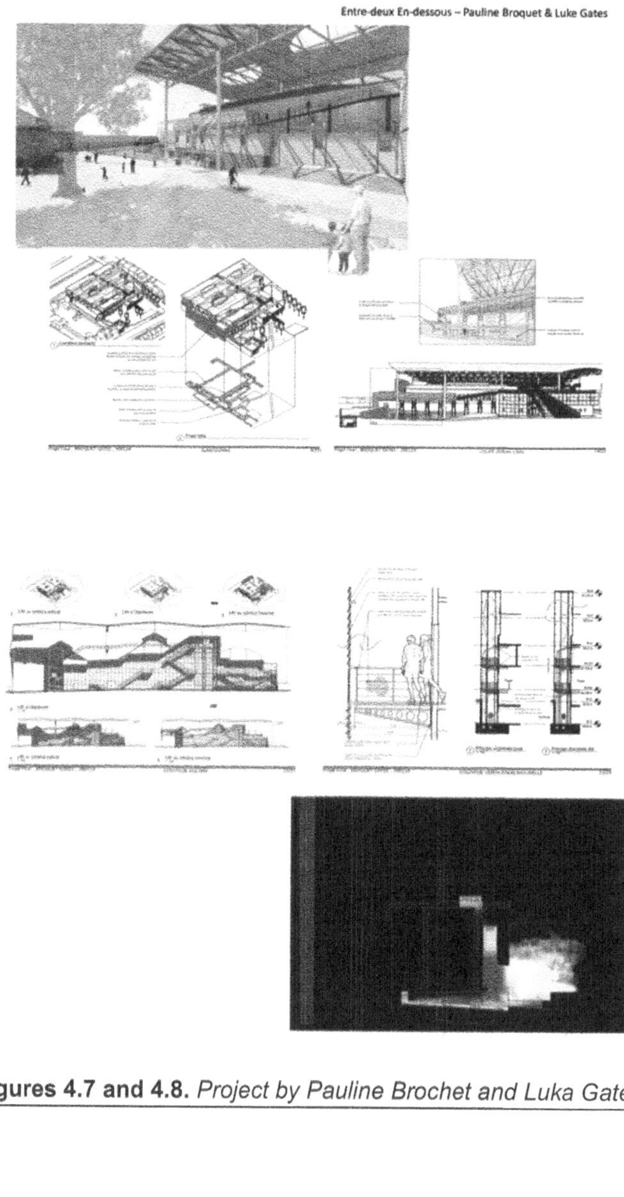

Figures 4.7 and 4.8. *Project by Pauline Brochet and Luka Gates*

4.5. Conclusion

Figure 4.9. *"An educational manifesto": works exhibition at* Fresnoy Panorama 21 *(2019)*

Let us outline a few elements to conclude.

Let us first highlight that the experience of this collaboration with an important institution was possible since it required a form of reciprocal familiarization and exchange of expertise: the attraction of architects towards a creation site and an exceptional building, as well as technical and "functional" expertise concerning a building whose users probably did not fully appreciate the importance and necessary updating.

Emphasizing this building's weaknesses took on a didactic dimension, both a "materialist" (what investment can we recommend to operate to meet which standards and what gains can be expected?), and a "conceptual" one (by what means? does the projectual equation work today and how does this echo the production issues of artists?).

When Hertzberger (2011) mentions this delicate balance to be achieved by the designer between *making space* and *leaving place*, it somehow designates the joint register of predictability and liberality (providing a formed but open space so that a large space is left for the freedom of uses) that architecture should engage in. While he will have tried to outline this program with the classic tools of project management in the second half of the 20th century, it seems to us that with the modeling–simulation couple, digital instrumentation precisely makes it possible to "push" further in this direction, a potential which some studies have noted. The control–appropriation dialectic therefore appears central to our approach to the project.

The climate equation is not confined to the logic of control, and the common is not satisfied with absolute freedom. Rather, this notional binomial finds a stimulating balance between projectual approaches, ranging from "under control" to "letting go".

But in some respects, the choice to work on this site will also have worked as a demonstration of the absurd. Here is an event-driven and iconic generation architecture of the 1980s, which has essentially ignored bioclimatic issues. Centered as it has been on the issues of design and image, it seems to remain stuck on positions at the antipodes of the polarities around which our thinking is organized today. Converting it without altering its posture[10] therefore raises its own kind of challenge. However, speculating about the transformation of this "heritage" work can in a way set a trend, due to its exceptional character and to the multiplied contradictions it forces to overcome, in order to grasp the stakes of the more ordinary rehabilitation of small-scale events inherited from the same period.

As has also been said, thinking about a third edition of *Le Fresnoy* on the very ground of what makes the "excellence" of this institution, namely the production of an image, represents very stimulating food for thought concerning the digital paradigm for designing the architectural project. We have a lot to learn from this cross-fertilization of expertise.

During the fall of 2019, in the form of an educational research manifesto, all of the work of the bachelor's degree group was the subject of an *in situ* demonstration, among the promotions of *Fresnoy*'s artists as part of their Annual panorama. This endeavor should have been renewed in the fall of 2020, if the Covid-19 pandemic had not interfered, and will in fact proceed in the fall of 2021 (see footnote 1).

---

10 Let us bear in mind that *Le Fresnoy* holds the "Remarkable Contemporary Architecture" label (previously the 20th century Heritage label).

With this implementation of the Tschumi project right in the middle of the artistic production of the building of which he is the author, we could almost have claimed a form of techno-aesthetic proselytism at the heart of the *Fresnoy* machine: yes, architecture is here among you, and feeds off your labors. Yes it is changing its nature!

But we must also mention the limitations of the experience, especially in year 1 (when the digital model had to be assembled from scratch, and delayed the project work accordingly). Thus, feedback was made difficult by the complexity of the site and the program. We could also wonder about the limitations of the means, methods and instruments available: any trip to the site is a financial challenge, the interdisciplinary approach is only made possible by the overinvestment of teachers working overtime, digital instrumentation is sometimes more straightforward, in addition to the level of teacher training which would undoubtedly need to be improved to further support this relationship.

New perspectives could be envisaged, starting with the establishment of propaedeutic work, a manual combining these technical, political, and aesthetic arguments. This implies problems and methodologies, calls for reference operations, and makes up a language, an appropriate lexicon, while reflecting a readership open to students, teachers, researchers, architects, project managers, inhabitants and amateurs. This is a possible way forward. There are others which would involve patiently going around these critical edifices while reflecting, pioneers as we are, about the actualization forms and conditions of these Utopian fragments without compromising their substance.

4.6. References

Banham, R. (1984). *The Architecture of Well-Tempered Environment*, 2nd edition. The Architectural Press, London.

Bosqué, C. (2016). La fabrication numérique personnelle, pratiques et discours d'un design diffus. Enquête au cœur des FabLabs, hackerspaces et makerspaces, de 2012 à 2015. PhD Thesis, Université Rennes 2, Rennes.

Citton, Y. (2014). *Pour une écologie de l'attention*. Le Seuil, Paris.

Coriat, B. (ed.) (2015). *Le retour des communs, la crise de l'idéologie propriétaire*. Les liens qui libèrent, Paris.

Dardot, P. and Laval, C. (2014). *Commun, essai sur la révolution au XXIème siècle*. La Découverte, Paris.

Daston, L. and Galison, P. (2012). *Objectivité*. Presses du réel, Dijon.

Dewey, J. (2010). *Le public et ses problèmes*. Gallimard, Paris.

Doutriaux, E. (2020). *Conditions d'air: Politique des architectures par l'ambiance.* Métis Presses, Geneva.

Ferre, A. (2005). *Conditioning.* Verb, Lidcombe.

Faces (ed.) (2005). *Air*, no. 60, Geneva.

Faces (ed.) (2006). *Energie*, no. 63, Geneva.

Faces (ed.) (2010). *Atmosphère*, no. 67, Geneva.

Gibson, J.J. (2014). *Approche écologique de la perception visuelle.* Dehors, Bellevaux.

Guiheux, A. and Tschumi, B. (eds) (1993). *Tschumi : une architecture en projet : Le Fresnoy.* Centre Pompidou, Paris.

Hertzberger H. (2011). *University Building in France, Nantes School of Architecture.* Holcim foundation for sustainable environment, Zürich.

Heschong, L. (1981). *Volupté thermique.* Parenthèses, Marseille.

Irigaray, L. (1983). *L'oubli de l'air – chez Martin Heidegger.* Editions de Minuit, Paris.

Lion, R., Mandoul, T., Fol, J., Hertwerck, F., Lefebvre, V. (2012). Urban planning and sustainable development. In *Climats*, Clément, G., Berdaguer, C., Scott, F.D., Boutang, Y.M., Gontier, P., Rouillard, D., Muller, H.W., Kaufmann, H., Picon, A., Gresset, P., Picon-Lefebvre, V., Sauerbruch, M., Vassal, J-P., Rey, E., Sobek, W., Lion, R., Mandoul, T., Fol, J., Hertwerck, F., Lefebvre, V. (eds). ENSA-Paris Malaquais et Infolio, Paris.

Oosterman, A. (ed.) (2013). *Volume #37: Is This Not a Pipe*? Stichting Archis, Amsterdam.

Oosterman, A. (ed.) (2015). *Volume 45#: Learning.* Archis, Amsterdam.

Sloterdijk, P. (2005). *Ecumes – Sphères III.* Hachette, Paris.

Tschumi, B. (1997). *Architecture in/of Motion.* Nai publishers, Rotterdam.

Tschumi B. (n.d.). *Le Fresnoy – studio national des arts contemporains.* Massimo Riposati, Paris.

Vassal, J-P. (2012). Ambiances et climats. In *Climats*, Clément, G., Berdaguer, C., Scott, F.D., Boutang, Y.M., Gontier, P., Rouillard, D., Muller, H.W., Kaufmann, H., Picon, A., Gresset, P., Picon-Lefebvre, V., Sauerbruch, M., Vassal, J-P., Rey, E., Sobek, W., Lion, R., Mandoul, T., Fol, J., Hertwerck, F., Lefebvre, V. (eds). ENSA-Paris Malaquais et Infolio, Paris.

Zask, J. (2011). *Participer – Essai sur les formes démocratiques de participation.* Le Bord de l'eau, Paris.

5

Drawing the Thickness of an Environment, Drawing the Thickness of Architecture

5.1. Introduction

What changes the study and representation of a struggling territory like the ZAD (*Zone à défendre*, Zone To Defend) of Notre-Dame-des-Landes? The purpose of this text is to present the experience of a group of students that was carried out on this territory, involving many concrete dimensions (political, ecological, etc.). This text aims, more precisely, to detail the consequences of this experience on a canonical problem of architecture: that of representation and so-called "technical" drawing. Indeed, it seems that the act of representation, often considered neutral, can, in reality, translate as a commitment and has a vital role in the communication of contemporary political and ecological issues. This text evokes some possible approaches in order to understand how to represent these issues, specific to this struggling territory, by means of drawing.

This text is a form of engagement in the experience of a territory in struggle: the ZAD of Notre-Dame-des-Landes. This formal register is the one that seemed to us the most appropriate to convey the experience of this place.

Chapter written by Christophe LAURENS, Valentin SANITAS, Clément GAILLARD, Tibo LABAT and Jérémie BUTTIN.

For a color version of all the figures in this chapter, see www.iste.co.uk/aithaddou/transition.zip.

5.2. Something is happening in Notre-Dame-des-Landes

The ZAD was founded to resist the Great West Airport Project and its worldview continues to resist the destruction of life. Constructions were made, crops were grown, a territory was conquered and a society created in relative autonomy and in accordance with this particular and ardently defended environment that is the bocage. How can architectural drawing account for what is being invented in the ZAD? This was the question posed by the project for the book *Notre-Dame des Landes ou le métier de vivre* (Notre-Dame des Landes or the profession of living); an educational project led by the DSAA *Alternatives Urbaines* at Vitry-sur-Seine. Since 2013, this team has been working on situations which call into question the urbanization modes that weaken the ecological and social fabrics of territories. In this context, the ZAD quickly emerged as a concrete expression of interesting alternatives for students. The project for surveying the cabins and their environment was organized in two stages: meeting and living. It was first necessary to present the idea to the ZAD members throughout several meetings. The goal was to work with surveys in order to accurately represent the constructions carried out in the ZAD within their environment. The detailed architectural statement also works here as the statement of an intimacy, something which requires the agreement and the participation of the inhabitants; one does not let a stranger enter one's home, least of all to take the measurements of the smallest pan. Proposing this idea of a book and circulating it made it possible to have the group of students become acquainted with the "Zadists", to welcome their imperatives and their advice, as well as to start making the list of constructions to be surveyed. The first exchanges quickly brought to light the intimate relationship between territory and inhabitants. These cabins, rich in their environment, in the history of their construction and that of their builders, demanded a particular drawn form. For this, a week was spent *in situ*, immersed in their collective life, to make the necessary drawings, sketches and architectural surveys; a week throughout which the moments of cooking or playing cards became as important as the working hours, so that the drawing was enriched by the vitality of the place. By paying attention to the living environment, and not merely to buildings, we tried to understand how the inhabitants of this ZAD call into question what it means to dwell in an environment.

5.3. Living in the bocage

Since the invention of the notion of biology in the 19th century, Jean-Baptiste Lamarck (1744–1829), who developed the notion of environment in France, proved that living beings necessarily interact with their environment; this enables their adaptation. By considering the bocage as their living environment, the "Zadists" adapt their planning arrangements, constructions and forms of life.

"Still, not every building is a dwelling." This famous quote from Heidegger is drawn from his conference in Darmstadt (1951), "Building, dwelling, thinking". Furthermore, he adds that "Bridges and hangars, stadiums and power stations are buildings, but not dwellings." This is the core of the fight carried out by those trying to dwell in the bocage. The relationship between dwelling and the environment is found in the integration of human activities with broader activities, here, the activities of the life of the bocage itself. If anything, what seems to be happening at Notre-Dame-des-Landes is this evolution of architecture towards an act of building embedded in the dwelling of a territory. "To build is in itself already to dwell," says Heidegger, because building is a way of incorporating the mesh of relationships which constitutes the environment, these interactions between plants, as described by Lamarck, but more broadly, between the worm and the earth, the bird and the clouds, the grasshopper, the cat and the human. Not only did the experiments at the ZAD of Notre-Dame-des-Landes recall the inseparable nature of these notions of environment and dwelling, but it enriched contemporary thinking about the environment. Often taken as an example, the act of building the ZAD appears as an ecology of relationships. The ZAD helps to manufacture the new cosmologies required to produce new alternatives. The big difference is that these new cosmologies are directly applied to a territory. Far from images and representation, the consideration of environments is born from the act of building itself. Just as the living must always adapt to their environment, constructions bear within them the traces of this consideration. The inventiveness that one might think is conceptual in understanding the living is actually constructive inventiveness.

5.4. Translating what we see

The question then arises on how to translate the inventiveness of this environment and of this building into action, inserting the human into the relational fabric of the bocage. How to account for what it means to live in the bocage of Notre-Dame-des-Landes? Cyrille Weiner, who has long photographed the links that small human communities have forged with their environment, made a first proposal through three photo notebooks. They gather around 30 images, which are not only fragments of a larger landscape, but also clues of a more extended living under the November sky. For our part, in order to imagine the drawings, we first had to live in the area for a week. The first choice was to live within the project. As we have shown, cabins cannot be reduced to constructions. Being full dwellings, we had to grasp the actions, habits and conflicts discords but also the crossroads, bushes, birds and other local specificities which weave this environment, and help them find their place in the drawings, as well as in the technical and constructive details.

5.5. Drawing architecture and the environment

The second significant choice was to use architecture tools, from on-site survey to computer-assisted drawing clean-up, in order to keep a descriptive look of these cabins which bear in them the traces of the bocage; a delicate balance between standards, disciplinary conventions and circumstantial inventions. It was the discussion of one of these standards which made it possible to understand the interplay of living things in the bocage. Following the moments of meeting and then dwelling came the moment of transcription. This educational project engaged the students in the organization of the project and in the field practices. Afterwards, a follow-up was put in place to get them involved in the editorial questions of the book. How to translate, by means of drawing software, the water or the clouds which are part, in this situation, of the actors having built these cabins? What depth of soil and length of roots can transcribe this very particular bocage? These and many other questions included in the survey protocol subsequently evolved during computer work in order to adapt the software to these unusual constructions. In-depth drawing involves finding, by means of architectural drawing, the ways to translate this singular relationship between habitat and environment. The choice of line thickness in architectural drawing is essentially related to visual hierarchy and must correspond to the conventional representation codes shared by a set of professions. When it is done correctly, architectural drawing makes it possible to classify elements; it is not simply a tool for representation, but constitutes in itself a record of the act of construction and arrangement of the elements. Line thickness, as well as the codes and conventional textures, help to support this hierarchy. These codes have been partially preserved in the plans, sections and elevations that were produced in the ZAD of Notre-Dame-des-Landes. The different line thicknesses are preserved, as well as the choice of differentiated textures. But the goal of thickness shifted from the line/object to the earth/environment: a section does not only show a house but all the beings and elements surrounding it. The cabins are located at an intermediate level between the depth of the ground and the high forest of the trees; most of the times, built with light foundations, the ZAD cabins appear as if they were placed on the ground. Thus, the function of the architectural drawing is preserved, since important elements, such as a fruit tree or a domesticated animal, are drawn and highlighted. These elements are not decoration added around the sections or the elevations of the different constructions, their function is not to make the architectural drawing more lively or illustrative. Some of them have an objective dimension, in so far as they designate a cabin's location; this is the case with hedges or trees. Other elements participate in the very life of the inhabitants, as much for pleasure as for agriculture or merely by chance.

5.6. The lessons of Geddes

Some drawn elements provide indications on the nature of the landscape. In addition to the architectural design of a building, there are the issues on how to represent an ecosystem. These issues were particularly studied by ecologists like Eugène Odum in the second half of the 20th century, but are actually older. It was Patrick Geddes (1854–1932), a biologist by training who became a sociologist and then an urban planner, who was the first to integrate ecosystem issues into geographic and urban representation. Geddes developed, at the beginning of the 20th century, what he called the "Valley Sections". These are sections showing the distribution of landscapes in relation to human activities; these sections connect a mountain, a relief up to a watercourse going through the hillsides. With these sections, Geddes intended to show the correlation between human activities and geographic location. The "valley sections" are most often without scale, but they are essentially functional representations which show the gradual passage from one landscape to another, and enable the different activities to be territorialized in various types of landscape (slope, coast, etc.).

The work initiated by Geddes in his valley sections hoped to combine the representation not only of habitat types and different human activities within their geographical situation, but also animal environments, a topical issue. While the drawings of the houses in the ZAD do not directly belong to the same register (they are not mere cut-sections across the entire bocage of Notre-Dame-des-Landes), the intention to develop a type of representation which is both architectural and ecosystemic is nonetheless present. Either in the choice of framing, the space left on the ground/in the air or in the bleed printing of all the drawings, there is this intention of working with that which escapes us, with everything outside the scope of architecture.

Beyond the different line thicknesses ensuring the legibility of the architectural drawings, there certainly is the thickness of the landscapes and the practices we had to undergo in order to try, with Nicolas Bouvier, to draw what he so aptly called The Way of the World.

5.7. References

Bouchain, P., Laurens, C., Weiner, C. (2018). *Notre-Dame-des-Landes ou le métier de vivre*. Loco, Paris.

Bouvier, N. (1963). *L'Usage du monde*. Droz, Geneva.

Canguilhem, G. (1952). *La Connaissance de la vie*. Hachette, Paris.

Geddes, P. (1915). *Cities in Evolution*. Williams & Norgate, London.

Heidegger, M. (1980). *Essais et conférences*, translated by André Préau. Gallimard, Paris.

PART 2

The Challenges of Ecological Transition Met by Research

6

Overall Comfort in Elementary Schools with Humid Tropical Climates: Methodology for Adapting Comfort Scales

Recent studies have shown that the design of school buildings today is inadequate to match children's needs. Indeed, the studies' results correlated school performance and health problems with a lack of comfort in classrooms.

This work addresses the issue of comfort in tropical environment school buildings. It raises questions at the heart of architectural thought, in the context of ecological and economic crises, and with the aim of providing a healthy and sustainable environment for children. It aims to suggest ways of improving the design of school buildings.

A student perception survey was distributed in parallel with the measurements of the interior rooms in elementary schools on Reunion Island. The analysis of various data made it possible to determine optimal comfort zones for students in different climatic zones of Reunion Island.

6.1. Introduction

The European and French standards currently applied are unsuitable for the child's physiology (ASHRAE Standard 2004). The child, who is a developing individual, is more sensitive to the environment than the adult. Furthermore, the

---

Chapter written by Magalie TÉCHER and Hassan AIT HADDOU.

For a color version of all the figures in this chapter, see www.iste.co.uk/aithaddou/transition.zip.

child's metabolism is more active for self-regulation and maintenance in a state of optimal comfort. In an educational setting, it is the environment where the school is set that must be adapted and comfortable so as to favor a learning disposition.

Several studies have shown that the design and quality of school buildings which acknowledge educational tools have a direct impact on the effectiveness, educational development and general concentration of children (Heschone Mahone Group 1999; Pilcher *et al.* 2002, p. 16; Vincent and Gissinger 2016, p. 30). Results have shown that suitable and comfortable premises, as well as a suitable pedagogy, have a positive impact on the education and success rate of children (Déoux and Girard 2010; M. Rodriguez *et al.* 2019).

On the other hand, in France, these standards are not appropriate for all types of climate. They originated in a national desire to standardize building design so as to ensure that interior environments comply with pre-established comfort requirements. In fact, they do not consider the different characteristics from one region to another nor the population's internal phenomena, which enable them to adapt to climatic particularities and remain in an optimal comfort situation.

This notion of "uniformity" is problematic in the case of humid tropical zones, such as Reunion Island, where climatic conditions do not make it possible to design in the same way as in a temperate zone. Indeed, a study carried out in Reunion highlighted that the share of energy consumption due to lack of comfort in non-residential buildings was 58% for air conditioning (Lenoir 2013).

The general goal of this work is to offer assistance in the design and rehabilitation of school buildings in a tropical climate context on Reunion Island, for which it will aim to:

– analyze the relationship between the sensations felt by children in classrooms and comfort scales;

– test the potential of using a questionnaire for the assessment of comfort scales adapted to children's needs;

– determine the optimal conditions of overall comfort in classrooms on Reunion Island, as inferred from the questionnaire.

6.2. General framework and methods

6.2.1. *Selection: choosing villages and elementary schools*

Reunion is an island located in the Indian Ocean, near Madagascar. Although it experiences the effects of a humid tropical climate, due to the steep and mountainous relief in its center, one can identify two distinct climatic zones (Figure 6.1): a "windward" zone in the east, which has a humid, windy and rainy climate, and a "leeward" zone in the west, where the climate is often sunny and dry (Garde *et al.* 2010).

Numerous studies (Kemajou *et al.* 2012, p. 12; Jindal 2019; Kumar *et al.* 2018) have highlighted the adaptations of populations to acclimatize to the particularities of their habitats. These particular traits have led populations to have optimal comfort requirements which are specific to their living environment. Due to the coexistence of several climatic zones on Reunion Island, we assume that there could be differences within the population: the population living in the West might have adapted to a hotter and drier climate than the population on the east of the island.

This work will attempt to verify this hypothesis by focusing on elementary schools (Jindal 2019; Kim and de Rear 2018) located in the coastal areas of Reunion Island, in order to assess the possibility of there being different comfort scales due to changing requirements, depending on the habitat.

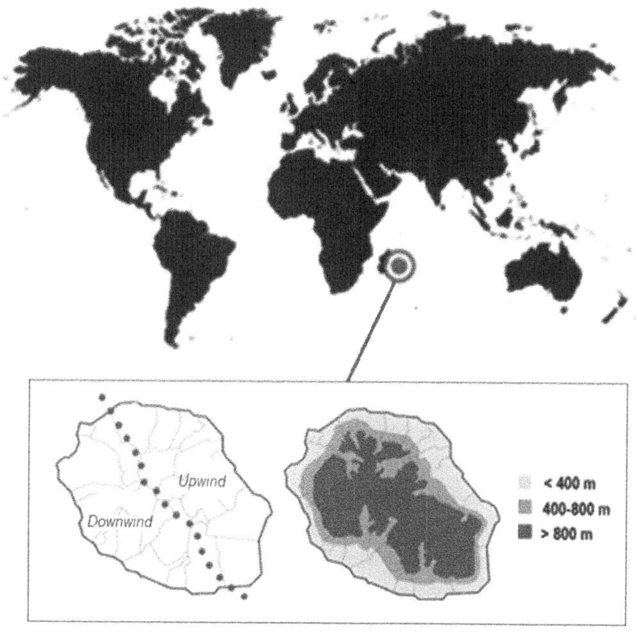

Figure 6.1. *The situation of Reunion Island and its particular traits*

The majority of the population is currently gathered within these coastal areas or "lowlands" (Figure 6.2), which is where the majority of schools are located.

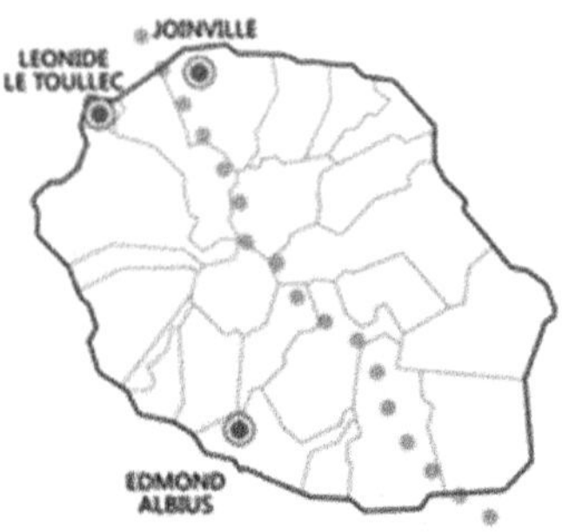

Figure 6.2. *Geographical location of schools*

Given their position, in order to be as exhaustive as possible and to better approach the reality of the island's microclimates, a choice was made to compare the schools located in the two main climatic zones (windward and leeward zones). The type of ventilation (natural ventilation being preferred in a majority of schools, we have focused our choice exclusively on this type of ventilation), the variation in the type of orientation and morphology of the buildings have been favored so as to extend the possibilities.

First, authorization requests were sent to the national education inspectors from the various sectors of the Reunion Island academy. Out of the four sectors of the island, the North, West and South sectors gave a positive response to a survey of this type within their schools. The eastern sector of the island was excluded from our study due to a negative feedback, reducing the study of the "windward" climatic zone to the northern sector of the island.

Secondly, we contacted all the school headteachers in these sectors to inform them about our approach. Since this study is based on the distribution of questionnaires and parameter measurement, it had to be carried out during the normal working hours in the classroom, and this is why only a small number of establishments accepted taking part in this survey.

School	Location	Construction period	Orientation	Highest level	Ventilation	Number of classes interviewed	Number of students interviewed	Recording day
Edmond Albius	Saint-Pierre	<1900	NW/SE	First floor	Natural	4	85	June 27, 2016
Léonide Letoullec	Le Port	1950-60	NNW/SSE	First floor	Natural	3	76	June 28, 2016
Joinville	Saint-Denis	1950-60	NW/SE	Ground floor	Natural	3	60	June 30, 2016

Edmond Albius *Léonide Letoullec* *Joinville*

Figure 6.3. *Characteristics and aerial views of schools (2016)*

The survey took place at three elementary schools in three towns on the island (Figures 6.2 and 6.3): the Joinville school in Saint-Denis (North), the Léonide Le Toullec school in the Port (West) and the Edmond Albius school in Saint-Pierre (South).

The classes were chosen so as to represent the different feelings that can be found within these schools in a homogeneous manner. Thus, all school levels, from CP to CM2, were represented within each school and as far as possible; the classes were located on separate floors and buildings, depending on the various orientations (Table 6.1).

School	Class	Orientation	Level	Sun protection system	Ventilation aid	Number of students interviewed
Edmond Albius	CP	NW/SE	Ground floor	Curtains	Roof fans	19
Edmond Albius	CE1	NNW/SSE	Ground floor	Curtains	Roof fans	24
Edmond Albius	CM1	NE/SW	Ground floor	Curtains	Roof fans	22
Edmond Albius	CM2	NW/SE	First floor	Curtains	Roof fans	20

Léonide Letoullec	CE1	NNW/SSE	Ground floor	None	Wall fans	25
Léonide Letoullec	CE2	NNW/SSE	First floor	Horizontal slat blinds NNW	Wall fans	24
Léonide Letoullec	CM1	NNW/SSE	First floor	Horizontal slat blinds NNW	Wall fans	27
Joinville	CE2	NW/SE	Ground floor	Curtains and roof overhangs NW	Wall fans	20
Joinville	CM1	NW/SE	Ground floor	Curtains and roof overhangs NW	Wall fans	22
Joinville	CM2	ENE/WSW	Ground floor	Curtains and roof overhangs WSW	Wall fans	18

Table 6.1. *Characteristics of the classrooms studied*

Taking all of these different factors into account reduced the number of classes studied accordingly.

6.2.2. *Setting up the questionnaire and collecting data*

In general, the questions aimed to reflect the students' level of satisfaction regarding a variety of comfort themes. Other questions targeted more open-ended research in order to understand in what ways students are active in their classroom environment (that is to say, whether they can improve their comfort in the classroom). Individual responses were collected for each student according to the following six categories:

– General information: the subject's gender, age, position in the class, and clothing (Clo). Clothing is important in the question of thermal comfort, in particular because it allows us to better understand the answers given by each person. Clothing values were adapted to the case study for a tropical climate (Saddok 2016).

– Overall comfort (Figure 6.4): the subject's satisfaction concerning several comfort themes, mainly thermal, light, acoustic and olfactory comfort, according to a satisfaction scale ranging from -3 to 3 (-3 very uncomfortable, -2 uncomfortable, -1 rather uncomfortable, 0 neutral, 1 rather comfortable, 2 comfortable and 3 very comfortable).

– Thermal comfort: questions the subjects about their thermal sensation at the precise moment of the questionnaire, knowing that this is a personal assessment, using the ASHRAE scale (Figure 6.4), a seven-point scale (-3 cold, -2 cool, -1 slightly cool, 0 comfortable, neutral, 1 slightly warm, 2 warm, 3 hot), a MC INTYRE three-point scale (Figure 6.4), that is, "At this precise moment I would like it to be: colder, the same, warmer" (Cantin *et al.* 2005).

– Visual comfort: determines the moments of visual discomfort in classrooms, the board's visibility, glare phenomena while executing the various visual tasks to be performed (looking at the board, reading, writing), depending on the time of the day. The goal of this study is to formulate new scales of comfort and to avoid situations that disrupt concentration and learning; therefore, we have chosen to highlight the annoying aspects that hinder comfort.

– Acoustic and olfactory comfort: describes the most annoying noises or smells when working.

– And finally, a free expression on points to be improved in their class, the solutions contemplated, overall satisfaction and a classification of overall comfort parameters, depending on their priority for work.

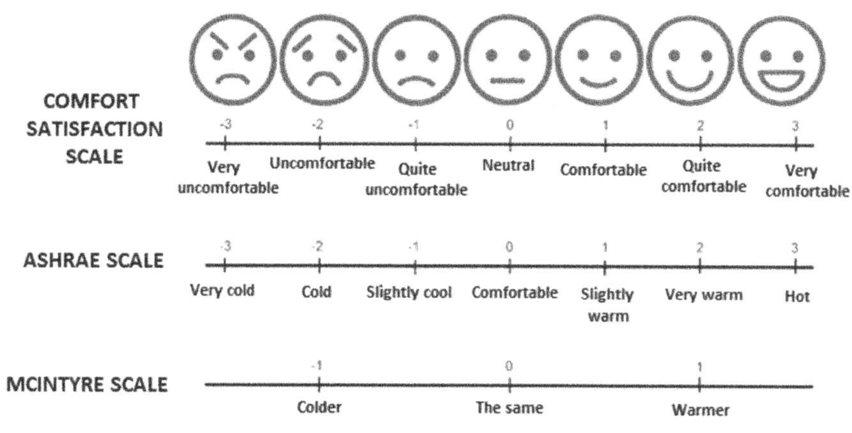

Figure 6.4. *Comfort scales implemented in the questionnaires*

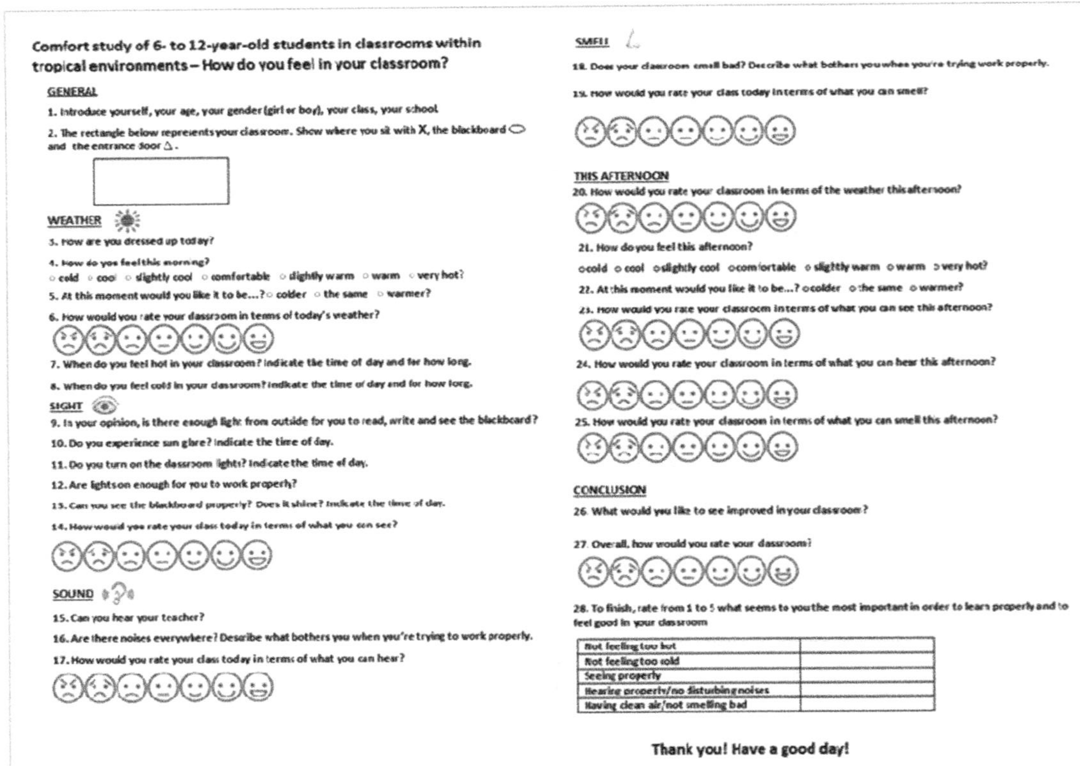

Comfort study of 6- to 12-year-old students in classrooms within tropical environments – How do you feel in your classroom?

GENERAL

1. Introduce yourself, your age, your gender (girl or boy), your class, your school
2. The rectangle below represents your classroom. Show where you sit with X, the blackboard ⬭ and the entrance door △.

WEATHER

3. How are you dressed up today?
4. How do you feel this morning?
○ cold ○ cool ○ slightly cool ○ comfortable ○ slightly warm ○ warm ○ very hot?
5. At this moment would you like it to be...? ○ colder ○ the same ○ warmer?
6. How would you rate your classroom in terms of today's weather?
7. When do you feel hot in your classroom? Indicate the time of day and for how long.
8. When do you feel cold in your classroom? Indicate the time of day and for how long.

SIGHT

9. In your opinion, is there enough light from outside for you to read, write and see the blackboard?
10. Do you experience sun glare? Indicate the time of day.
11. Do you turn on the classroom lights? Indicate the time of day.
12. Are lights on enough for you to work properly?
13. Can you see the blackboard properly? Does it shine? Indicate the time of day.
14. How would you rate your class today in terms of what you can see?

SOUND

15. Can you hear your teacher?
16. Are there noises everywhere? Describe what bothers you when you're trying to work properly.
17. How would you rate your class today in terms of what you can hear?

SMELL

18. Does your classroom smell bad? Describe what bothers you when you're trying work properly.
19. How would you rate your class today in terms of what you can smell?

THIS AFTERNOON

20. How would you rate your classroom in terms of the weather this afternoon?
21. How do you feel this afternoon?
○ cold ○ cool ○ slightly cool ○ comfortable ○ slightly warm ○ warm ○ very hot?
22. At this moment would you like it to be...? ○ colder ○ the same ○ warmer?
23. How would you rate your classroom in terms of what you can see this afternoon?
24. How would you rate your classroom in terms of what you can hear this afternoon?
25. How would you rate your classroom in terms of what you can smell this afternoon?

CONCLUSION

26. What would you like to see improved in your classroom?
27. Overall, how would you rate your classroom?
28. To finish, rate from 1 to 5 what seems to you the most important in order to learn properly and to feel good in your classroom

Not feeling too hot	
Not feeling too cold	
Seeing properly	
Hearing properly/no disturbing noises	
Having clean air/not smelling bad	

Thank you! Have a good day!

Figure 6.5. *Copy of the questionnaire*

6.2.3. *Overall comfort measurement*

In order to synchronize the measurements of class environments with student satisfaction, room parameters were measured at a height of 0.6 m on an unoccupied working table, which corresponds to a seated person according to the ISO 7726 standard[1] relating to the ergonomics of thermal environments.

The temperature of the wall could not be measured due to equipment-related reasons, this parameter was excluded and as a consequence, the operating temperature was not taken into consideration. The measured parameters are as follows:

– Air temperature, humidity and CO_2 concentration measured with a Voltcraft C0-100 ambient air meter (manufactured by the German company Conrad), which performs air temperature readings from 0 to 50°C, relative humidity from 20 to 90% and the CO_2 rate from 0 to 3000 ppm, with a ± 5% accuracy.

– The air speed was measured over a period of one minute by a Testo 410-1 Anemometer (manufactured by the German company Testo SE & Co. KGaA), which performs air speed readings from 0 to 20 m/s with an accuracy of ±0.2 m/s.

– While carrying out the questionnaire, lighting was measured with and without the lights on, using a Testo 540 Luxmeter (manufactured by the German company Testo SE & Co. KGaA), which performs lighting readings from 0 to 99999 Lux with an accuracy of ±3 Lux.

– The acoustic level was measured using an IDB ACOEM acoustic level meter (manufactured by the French company ACOEM), which records noise levels ranging from 30 to 120 dB in two ranges, with an accuracy of ±1 dB.

6.2.4. *Session planning*

The questionnaires and room parameter readings were carried out in one-day sessions, from June 27th to 30th, 2016, for each school.

The same protocol was followed in each visited classroom:

– Once in the classroom, the measuring devices were set up at two points: at the point furthest from the windows (in the middle of the classroom) and at the point closest to the windows.

---

1 ISO 7726:1998 defines the minimum characteristics of devices, as well as the methods for measuring the physical quantities of an environmental setting.

– The study's purpose was explained as the questionnaires were handed out, a first reading was carried out in order to answer any questions from the students.

– While the students filled out their questionnaires, the room's physical parameters were recorded.

– Two measurement sessions were taken for each point at an interval of 20–30 minutes for each class.

– The students filled out the questionnaires twice, once in the morning and once in the afternoon, which produced a total of four measurements of the same parameter for the same class, and two sets of answers for each person questioned.

6.2.5. *Sample profile*

The subjects of this study are students aged 6 to 12, questioned in their classrooms. In parallel, some teachers also answered the questionnaire, which made it possible to deepen knowledge about the classroom's environment.

In total, 221 subjects (Table 6.2), including teachers, took part in this field study, which took place during the southern winter: 85 students participated at Edmond Albius school in BoisD'Olives (Saint-Pierre), 76 students at Léonide Le Toullec school in the city of Port and 60 students at Joinville school in Saint-Denis.

	Edmond Albius	Leonide Letoullec	Joinville
Girls	50	50	29
Boys	35	26	31
Mean age	8.91	9.01	9.779
Girl's clothing mean	0.588	0.493	0.576
Boy's clothing mean	0.457	0.487	0.485
Window's clothing mean	0.529	0.482	0.558
Environment's clothing mean	0.542	0.5	0.532

Table 6.2. *Profile of the study sample [gender-age-clothing-positioning] (mean values)*

6.2.6. *Methodology*

Data processing required several steps (Figure 6.6), including data classification. Each response was coded in algebraic form and processed as quantitative data.

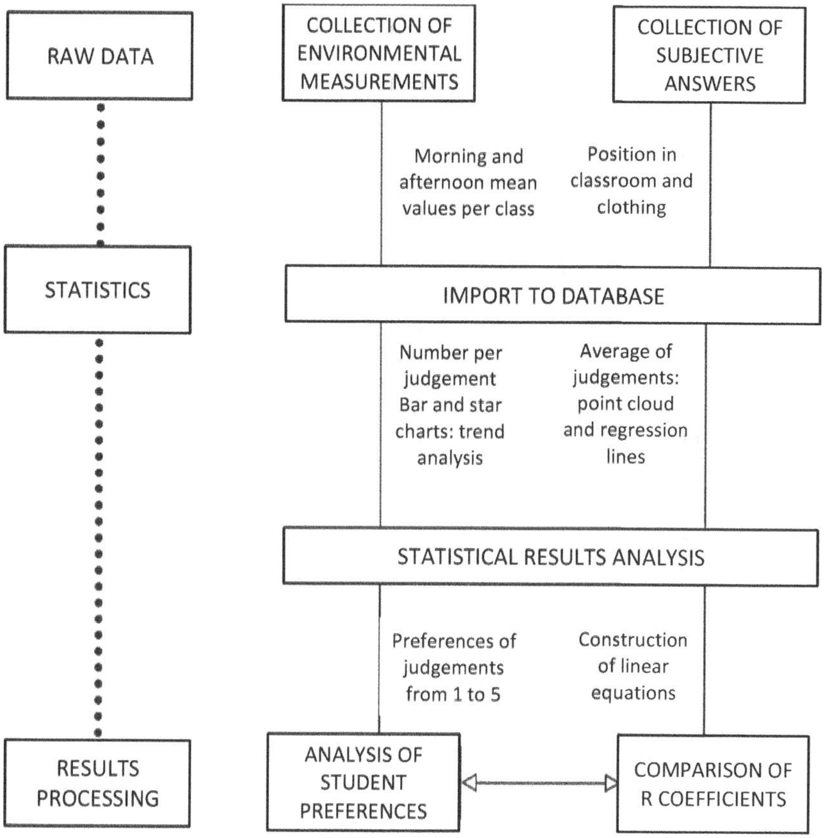

Figure 6.6. *Data classification and analysis process*

All of the data collected were organized in a tree with multiple branches, in order to better understand the type of response provided. The subjects were grouped following several categories depending on sex, age, clothing, positioning.

Mean satisfaction per group was then calculated in order to:

– Present the mean satisfaction per group depending on the position in the classroom. These groups are made up of students who answered the questionnaire in the same environmental conditions, in other words, they were groups belonging to the same class.

– Carry out the linear regression analysis, that is, to analyze the relationship between the means of these satisfaction indicators (satisfaction levels analyzed on their own are integers, but their mean is not necessarily the same: it is the ratio of the sum of the satisfaction levels of the same group by the number of students in the group) according to the measurements obtained in the classrooms.

6.3. Results and discussions

6.3.1. *Satisfaction analysis*

6.3.1.1. *Thermal comfort*

From the answers given to the questions on thermal comfort, we calculated the student's thermal satisfaction for each class, in the morning and in the afternoon.

Thermal perceptual satisfaction levels in the morning (Figure 6.7), show that the students mostly expressed that they were comfortable, with satisfaction means measuring S (A)[2] = 1.09, S (J)[3] = 1.58, S (L)[4] = 1.03. Overall a large number of classrooms qualified as being in a very comfortable situation. In the afternoon (Figure 6.7), the results show a slight decrease in overall satisfaction S (A) = 1.02, S (J) = 0.25, S (L) = 1. As a matter of fact, during the day, the air temperature increased while the relative humidity decreased, which may explain the decrease in general satisfaction.

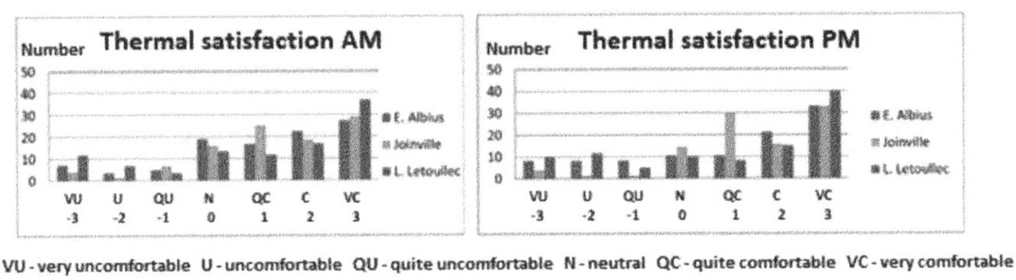

Figure 6.7. *Thermal satisfaction (morning and afternoon)*

The perceptual satisfaction levels of the ASHRAE scale in the morning (Figure 6.8) shows that the majority of students stated that they were comfortable S (A) = -0.235, S (J) = -0.079, S (L) = -0.3. The result is very close to zero, with a slight negative value which shows that the students felt cold rather than hot, whereas

2 S (A) = Mean satisfaction at Edmond Albius – Saint-Pierre school.

3 S (J) = Mean satisfaction at Joinville – Saint-Denis school.

4 S (L) = Mean satisfaction at Leonide Letoullec – Le Port school.

in the afternoon (Figure 6.8) the perceptual satisfactions of the ASHRAE scale show that the students mostly stated that they were comfortable, with a higher proportion feeling hot rather than cold. S (A) = 1.4, S (J) = 0.68, S (L) = 0.6.

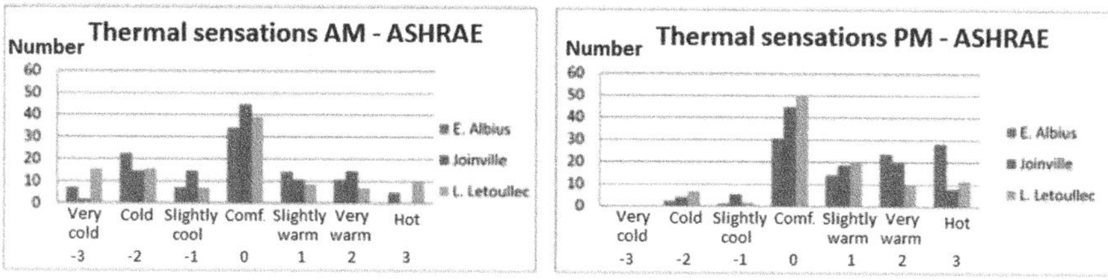

Figure 6.8. *ASHRAE Thermal satisfaction (morning and afternoon)*

The perceptual satisfactions of the MCINTYRE scale (Figure 6.9) in the morning show that the majority of students express the fact that they do not want any change and want the situation to stay the same S (A) = 0.105, S (J) = -0.13, S (L) = 0.13. In the afternoon, the perceptual satisfaction levels show that the students mostly state that they do not want any change. Thermal comfort parameters remain the same, with a larger number of students wishing that it was colder rather than hot: S (A) = -0.18, S (J) = -0.32, S (L) = -0.12.

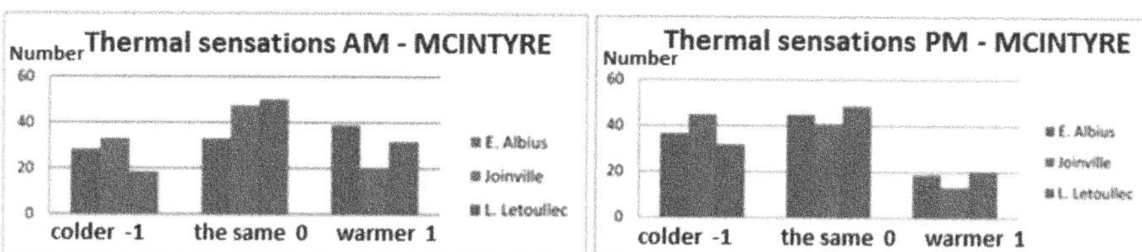

Figure 6.9. *MCINTYRE Thermal satisfaction (morning and afternoon)*

6.3.1.2. *Visual comfort*

The perceptual satisfaction levels relating to light comfort (Figure 6.10) in the morning show that the students express great satisfaction. This phenomenon can be explained by the fact that in the morning soft light enters directly through the window and does not cause glare phenomena, the brightness is constant outside and the curtains are mostly open S (A) = 1.43, S (J) = 1.66, S (L) = 1.55. In the afternoon, perceptual satisfaction levels showed that the students mostly express satisfaction with the lighting parameters.

Actually, these survey days were very bright days with very clear skies and few clouds. S (A) = 1.32, S (J) = 1.71, S (L) = 1.8, the values are therefore slightly higher than in the morning because the classrooms are more illuminated.

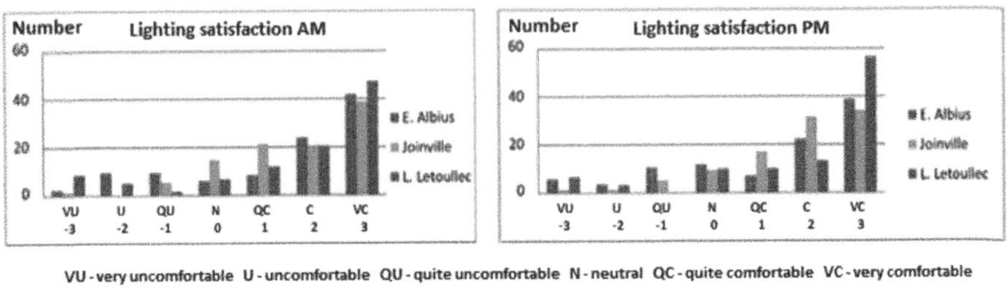

Figure 6.10. *Visual satisfaction (morning and afternoon)*

6.3.1.3. *Acoustic comfort*

The perceptual satisfaction levels relating to acoustic comfort (Figure 6.11) in the morning show that Léonide Le Toullec's students expressed great satisfaction because the classes are far from the street, and the only disturbing noises come from the classroom itself and the neighboring classes. The students of Edmond Albius expressed great dissatisfaction, due to the very close presence of a busy road and open windows. The students of Joinville reflected average results, since the classes are located on a road in the city center where traffic speed is reduced. This is achieved by reducing disturbing noises and leaving the windows overlooking the street closed. S (A) = -1.28, S (J) = 1.065, S (L) = 2.05.

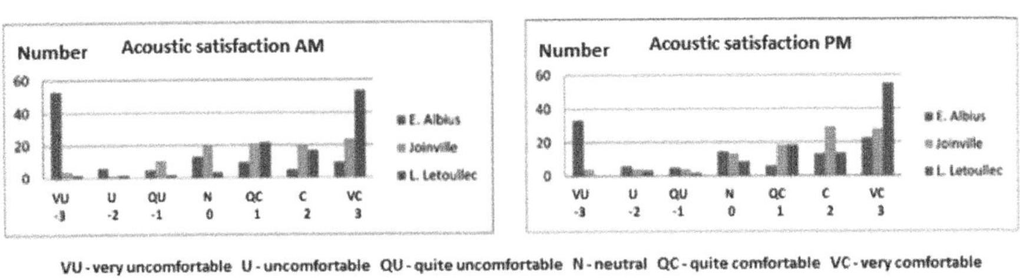

Figure 6.11. *Acoustic satisfaction (morning and afternoon)*

In the afternoon, the perceptual satisfaction of acoustic comfort remained constant at Léonide Le Toullec. The students from Edmond Albius expressed less dissatisfaction with the presence of disturbing noises coming from the busy street than the students in the morning, and this is perhaps due to a certain habituation to the noise during the day. Finally, the students from Joinville also reflected average results, feeling rather comfortable. S (A) = -0.2, S (J) = 1.35, S (L) = 2.01.

6.3.1.4. *Olfactory comfort*

In general, the perceptual satisfaction of olfactory comfort (Figure 6.12) in the morning showed that students expressed great satisfaction with olfactory sensations in their class S (A) = 0.47, S (J) = 1.42, S (L) = 0.78. In the afternoon, the perceptual satisfactions of olfactory comfort showed that the students always expressed a great satisfaction with the olfactory sensations in their class S (A) = 1.72, S (J) = 1.84, S (L) = 1.4.

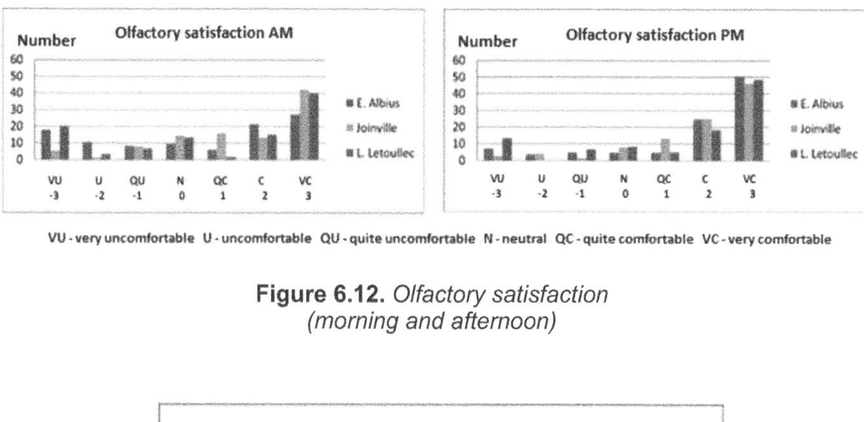

VU - very uncomfortable U - uncomfortable QU - quite uncomfortable N - neutral QC - quite comfortable VC - very comfortable

Figure 6.12. *Olfactory satisfaction (morning and afternoon)*

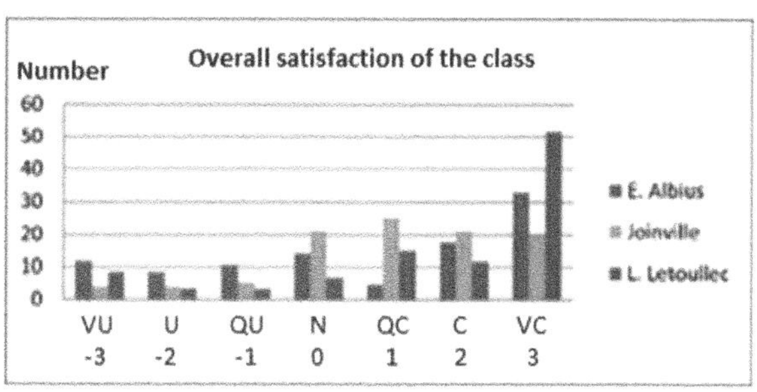

VU - very uncomfortable U - uncomfortable QU - quite uncomfortable N - neutral QC - quite comfortable VC - very comfortable

Figure 6.13. *Overall satisfaction*

6.3.1.5. *Overall comfort*

The mean perceptual satisfaction levels of the classes (Figure 6.13) generally showed that the students expressed satisfaction with their classrooms, and in particular, that Léonide Le Toullec school students expressed great satisfaction with their classroom. Yet, we must consider the possibility they may have interpreted the question "do you like your class?" in terms of yes/no, rather than answering the deeper question "how would you judge your class in view of each of the comforts levels we have previously described?" S (A) = 0.76, S (J) = 1.01, S (L) = 1.58.

6.3.2. *Linear regression analysis*

These regressions were carried out according to the environmental indicators (room parameters) measured:

– for the study of thermal comfort, the mean satisfaction levels perceived by the children observed as a function of the air temperature aT (in °C) and of the relative humidity rH (in %);

– for lighting comfort, mean satisfaction as a function of lighting (in Lux);

– for acoustic comfort, the mean satisfaction as a function of the acoustic level (in dB(A));

– for olfactory comfort, the mean satisfactions according to the level of CO_2 (in PPM).

Regression analyses allow us to obtain linear equations (type Y = aX + b) and regression coefficients, which are variables between 0 and 1 that enable us to express the relationship between two types of variables, X and Y. Thus, the greater the coefficient and the closer to 1, the greater the relationship between the two elements, that is to say, the mathematical model is validated. In fact, it is capable of precisely determining a large number of points, which can be exploited.

For this study, we considered that a regression coefficient greater than 0.5 demonstrates a relationship sufficiently close to the elements X (environmental indicators) and Y (satisfaction) to permit an exploitable prediction.

6.3.2.1. *Thermal comfort*

A first linear regression line as a function of the air temperature aT is obtained for each school and is therefore adapted to each village.

The R coefficients obtained (Figure 6.14) thanks to these regression lines as a function of the air temperature for the three schools are greater than 0.7734. This proves a direct and readable relationship between air temperature and the thermal satisfaction levels of the children in the classroom.

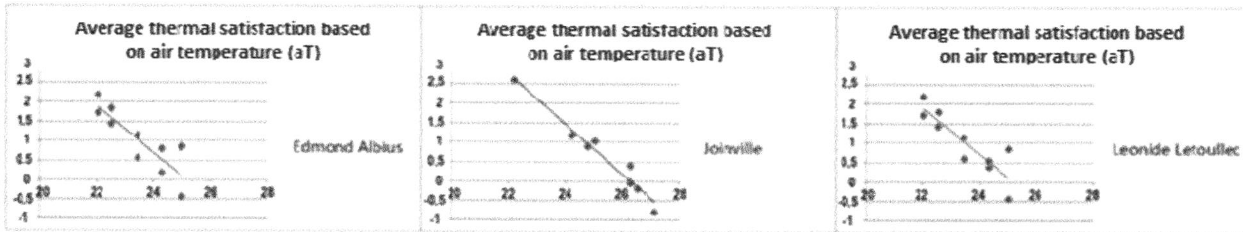

Indicator	THERMAL SATISFACTION			
	School	E. Albius	Joinville	L. Letoullec
	Town	Saint-Pierre	Saint-Denis	Le Port
Based on aT	Equation	y = -0.6017x+15.148	y = -0.6395x+16.799	y = -0.6013x+15.134
	R =	0.8144	0.9577	0.7734

Figure 6.14. *Lines, equations and regression coefficients obtained (air temperature)*

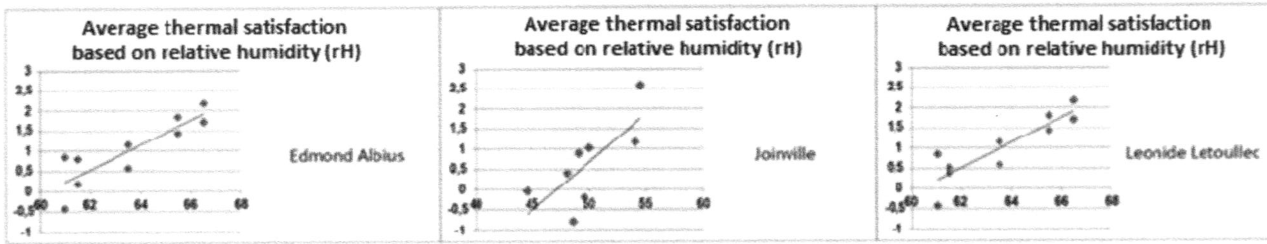

Indicator		THERMAL SATISFACTION		
	School	E. Albius	Joinville	L. Letoullec
	Town	Saint-Pierre	Saint-Denis	Le Port
Based on rH	Equation	y = 0.3045x-18.342	y = 0.2384x-11.232	y = 0.3047x-18.362
	R =	0.7516	0.565	0.782

Figure 6.15. *Lines, equations and regression coefficients obtained (relative humidity)*

A second regression line of thermal satisfaction levels as a function of the relative humidity rate was obtained (Figure 6.15). The regression coefficients obtained (> 0.565) thanks to these graphs proved the existence of a relationship between the relative humidity rate and the thermal satisfaction levels of children.

This method is applicable for the determination of thermal comfort zones with air temperature and relative humidity.

6.3.2.2. *Visual comfort*

The regression lines of the lighting satisfaction levels obtained (Figure 6.16), depending on the lighting in the classrooms, made it possible to obtain very large regression coefficients (significant correlation), very close to 1 (> 0.9).

We can therefore confirm the almost linear relation between lighting and children's light satisfaction. This model is validated for lighting comfort.

6.3.2.3. *Acoustic comfort*

The regression analysis of acoustic satisfaction as a function of the acoustic level recorded in the classrooms revealed very low regression coefficients for two schools (Figure 6.17), the relationship between acoustic satisfaction and the acoustic level cannot be demonstrated in this way, with the coefficients being low (< 0.2). The relationship is not significant. These results may be rooted in the difficulty to maintain similar conditions from one class to another (open/closed windows/doors, fans on/off, etc.).

6.3.2.4. *Olfactory comfort*

The analysis of the three regression lines for olfactory satisfaction levels (Figure 6.18) as a function of the CO_2 rate made it possible to obtain fairly high regression coefficients (> 0.6224), thus highlighting the relationship between the level of CO_2 and student olfactory (dis)satisfaction. This method is applicable for the study of olfactory comfort for determining comfort scales.

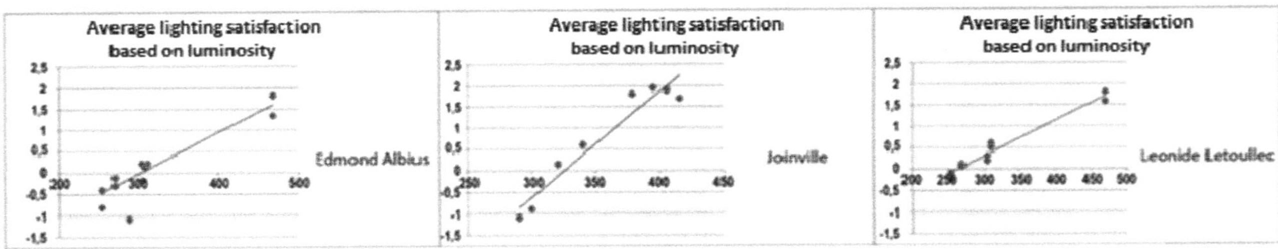

Indicator	LIGHTING SATISFACTION			
	School	E. Albius	Joinville	L. Letoullec
	Town	Saint-Pierre	Saint-Denis	Le Port
Based on luminosity	Equation	y = 0.0128x-3,37	y = 0.0249x-8.0734	y = 0.0084x-2.2194
	R =	0.9227	0.9278	0.9716

Figure 6.16. *Lines, equations and regression coefficients obtained (lighting)*

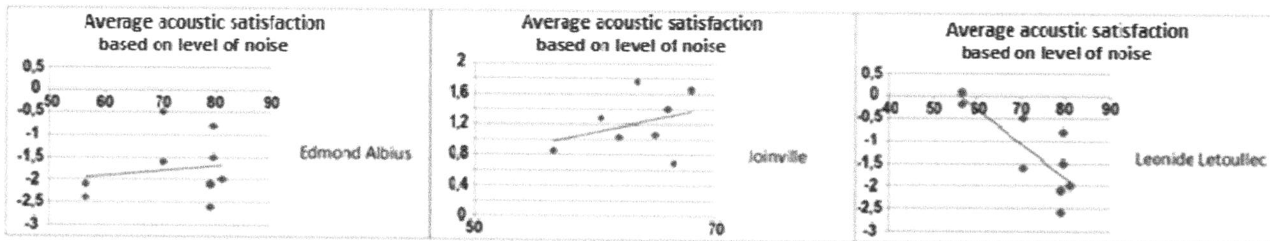

Indicator	ACOUSTIC SATISFACTION			
	School	E. Albius	Joinville	L. Letoullec
	Town	Saint-Pierre	Saint-Denis	Le Port
Based on acoustic lev.	Equation	y = 0.0113x-2.5918	y = 0.0336x-0.9194	y = -0.07633x+4.2704
	R =	0.0263	0.1146	0.6559

Figure 6.17. *Lines, equations and regression coefficients (acoustic level)*

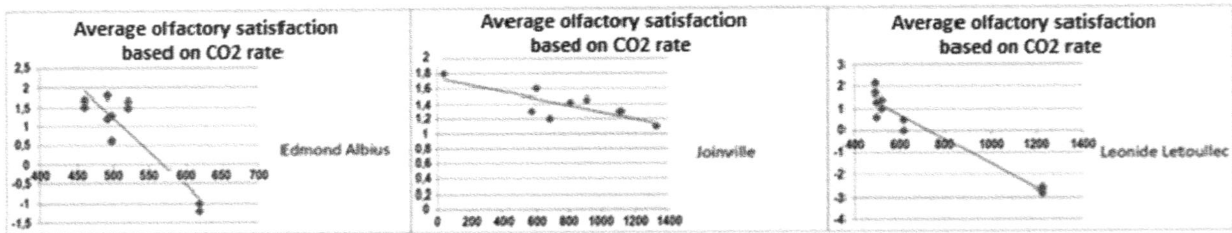

Indicator	OLFACTORY SATISFACTION			
	School	E. Albius	Joinville	L. Letoullec
	Town	Saint-Pierre	Saint-Denis	Le Port
Based on CO2 rate	Equation	y = -0.0177x+10.041	y = -0.0005x+1.7426	y = -0.056x+4.059
	R =	0.8172	0.6224	0.9306

Figure 6.18. *Lines, equations and regression coefficients obtained (CO2 rate)*

6.3.3. *Determination of global comfort scales*

The satisfaction scale (from -3 very uncomfortable to 3 very comfortable) used in this study made it possible to calculate student satisfaction mean values for each class and each school.

For this study, the satisfaction scales between 1 (rather comfortable) and 3 (very comfortable) determine optimal satisfaction with respect to the thermal, visual, acoustic and olfactory environment. They represent the extremes of the acceptable comfort zone and make it possible to determine new comfort zones over time. Hypothetically, these comfort zones – determined on the basis of these evaluation scales – are those which satisfy classroom occupants the most.

In a sense, the linear regression equations obtained are capable of predicting occupant satisfaction as a function of an environmental indicator (room parameter). In the equation $Y = aX + b$, variable Y represents occupant satisfaction, while variable X serves as a quantifiable environmental indicator.

In this way, it is possible to obtain the limits of our comfort zones by solving the linear regression equations with $Y = 1$ then $Y = 3$.

6.3.3.1. *Thermal comfort*

Solving the equations made it possible to define more optimal thermal comfort scales (Table 6.3). In point of fact, two large zones were identified. In our study, we classified the lowlands of Reunion Island into two zones: the windward zone (including Saint-Denis) and the leeward zone (including Saint-Pierre and Le Port).

	THERMAL COMFORT			
	School	E. Albius	Joinville	L. Letoullec
	Town	Saint-Pierre	Saint-Denis	Le Port
Thermal acceptability	aT	20.19 < aT < 23.51	21.6 < aT < 24.7	20.17 < aT < 23.50
	rH to aT Max	63.5 % at 23.51°C	51.3 % at 24.7°C	63.5 % at 23.50°C
	rH to aT Min	70 % at 20.19°C	59.7 % at 21.6°C	70.1 % at 20.17°C

Table 6.3. *Thermal comfort scales (air temperature and relative humidity)*

The table shows very similar air temperature (aT) and relative humidity (rH) scales for the cities of Saint-Pierre and Le Port: this confirms the hypothesis of the same thermal zone including these two cities, and the individuals from these two territorial zones deemed it acceptable to be included under the same thermal comfort zone.

On the contrary, the city of Saint-Denis presents different results. In fact, the air temperature scales are significantly higher. As the school is located in the city center of Saint-Denis, there is an urban heat island effect (Malet 2017). The population questioned presents an adaptation to the climate whose temperatures are higher; they have greater resistance when the air temperature increases. However, relative humidity scales are lower compared to the two other cities. The windward zone is characterized by high relative humidity percentages compared to the other zone (around 80% all year round). These results show that the population has adapted to higher temperatures, but that the relative humidity level outside must be reduced significantly to gain comfort inside the building.

A comparison was carried out between the thermal comfort zone that could be found using the ASHRAE scale (Table 6.4)), with the results found using the previous thermal satisfaction scale.

	THERMAL COMFORT			
	School	E. Albius	Joinville	L. Letoullec
	Town	Saint-Pierre	Saint-Denis	Le Port
ASHRAE thermal acceptability	aT	21,9 < aT < 24,3	20,6 < aT < 27,8	22,5 < aT < 25
	rH to aT Max	61,9 % at 24.3°C	44.7 % at 27.8°C	60 % at 25°C
	rH to aT Min	66,7 % at 21,9°C	60,9 % at 20,6°C	65 % at 22,5°C

Table 6.4. *ASHRAE Thermal comfort scales (air temperature and relative humidity)*

Overall, the results obtained by the two scales are consistent, and the difference is small. The scales found for the towns of Saint-Pierre and Le Port present a similarity as they belong to the leeward area, and for the town of Saint-Denis, the temperature results are higher and the relative humidity is lower.

6.3.3.2. *Visual comfort*

Standard NF EN 12464-114[5] recommends lighting levels of 300 lux for classrooms and 500 lux above the blackboard. The results found by solving the equations (Table 6.5) prove that the students from Reunion Island feel more comfortable when lighting levels are greater than those provided for by the standard. Indeed, the minimum values found and considered to be acceptable are slightly higher than the standard values.

5 Standard NF EN 12464-1 prescribes the lighting requirements for people in indoor workplaces and those that will meet the needs of visual comfort and visual performance.

	VISUAL COMFORT			
	School	E. Albius	Joinville	L. Letoullec
	Town	Saint-Pierre	Saint-Denis	Le Port
Visual acceptability	Lum (lux)	338.8 < Lum < 493.8	364.4 < Lum < 444.71	383.3 < Lum < 621.4

Table 6.5. *Visual comfort scales (lighting)*

These results prove that children need brightly lit classrooms to learn and work properly. There is significant work to be done on natural lighting, an important and very present resource on Reunion Island, in order to reduce artificial lighting and the building's energy consumption.

6.3.3.3. *Olfactory comfort*

Solving the equations (Table 6.6) in order to obtain optimal scales of olfactory comfort did not provide consistent results. In fact, the results found from one school to another often differed too much and seemed too inconsistent to be interpreted as a whole.

	OLFACTORY COMFORT			
	School	E. Albius	Joinville	L. Letoullec
	Town	Saint-Pierre	Saint-Denis	Le Port
Olfactory acceptability	CO_2 rate	$339.56 < CO_2 < 505$	$-2154.8 < CO_2 < 1485.2$	$18.9 < CO_2 < 54.6$

Table 6.6. *Olfactory comfort scales (CO2 rate)*

Scales of olfactory comfort cannot be determined in this study. As for the study of acoustic comfort, it is impossible to validate this model for studying olfactory comfort, another method should be used.

6.4. Conclusion

The method chosen for this study made it possible to determine comfort scales according to the comfort level of individuals. On the one hand (Figure 6.19), the measured comfort temperatures are close to those presented in the literature, for example, that neutral temperature favors better performance (Pilcher *et al.* 2002). On the other hand, the comfort scales measuring relative humidity rates are narrower than those defined as comfortable relative humidity rates (Becker *et al.* 2007, p. 16).

As a first approximation, we can say that elementary school students in Réunion seemed to prefer cooler and more humid environments compared to people living in temperate or cold climates. Reunionese do not seem to experience thermal discomfort in an indoor air temperature range between 27 and 28°C, while these temperatures are considered too hot by the international standard, as shown by studies carried out in Cameroon (Kemajou *et al.* 2012), India (Jindal 2019; Kumar *et al.* 2018) and Australia (Kim and de Rear. 2018).

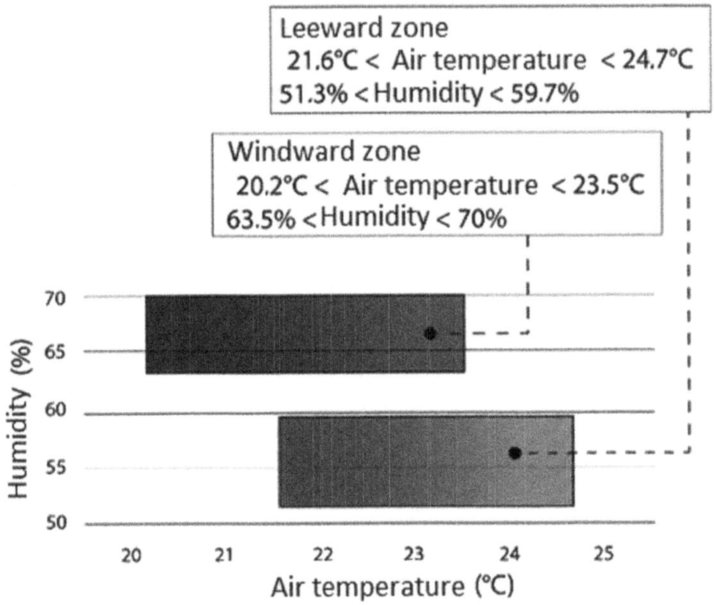

Figure 6.19. *Thermal comfort scales – graphic representation*

Likewise, the comfortable indoor air temperature results found for the two schools located in the windward area are similar, but significantly different from those found in the school located in the leeward area. This seems to confirm the hypothesis of the presence of "adaptive" phenomena in each of the populations living in each climatic zone, thus confirming the importance of a contextualized definition of comfort. The results show that subjects living in the hotter and drier zone prefer, on the contrary, cooler and more humid environments, while subjects living in the cooler and more humid zone prefer warmer and drier environments.

Finally, the comfortable lighting levels (Figure 6.20) found are higher than those predicted by the standards and regulations (*Association Française de l'Eclairage*, French Lighting Association, 1987). This result confirms the importance of having a contextualized comfort scale, potentially explained by an adaptive phenomenon of the subjects living in tropical zones and who evolve in places with brighter atmospheres than in temperate zones.

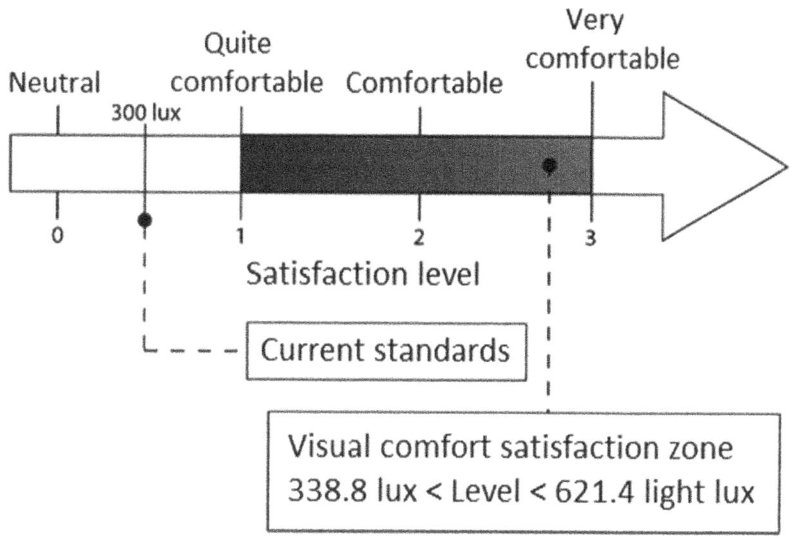

Figure 6.20. *Visual comfort scales – graphic representation*

However, as the study is restricted to a specific climatic context, and more particularly, to the windward and leeward areas of Reunion, these results cannot be applied in other contexts. On the other hand, the study was carried out with a limited sample of individuals and it was difficult to maintain the same survey conditions for each case study. Therefore, the results should be interpreted as general trends in elementary schools in Reunion.

The goal was to propose new ergonomic physical indicator scales for measuring overall comfort, some of which could not be measured (wall temperature, operating temperature, type of sound sources, etc.).

The comfort definition models used in the approach of this study did not make it possible to find acceptable scales of acoustic and olfactory comfort. It is important to pursue and deepen the models in the literature, which would make it possible to define more suitable comfort ranges. Also, a better relationship between satisfaction and comfort could be offered (Jindal 2019; Kim and de Rear 2018; Kumar *et al.* 2018).

In addition, it would be interesting to change the populations between schools, so as to measure their feelings in the same way. It is possible that living in a climatic context for a certain time modifies our perception by a compensation phenomenon in response to an uncomfortable situation.

Finally, this study could also be extended to a greater number of cases of school buildings, and to buildings with different occupations, in order to validate the hypothesis of the "adaptive" phenomena of the young population of Reunion.

6.5. Acknowledgments

This research work, carried out as part of a Master's internship, is an independent work. The author would like to thank Hassan Ait Haddou for helping in the development of this work, as well as the academy inspectors for their authorizations. Special thanks to the students, educational teams and headteachers who took part in this study.

6.6. References

ASHRAE (2015). Thermal environmental conditions for human occupancy. *ASHRAE*, 54–55.

Association Française de l'Eclairage (1987). Recommandations relatives à l'éclairage des locaux scolaires. *LUX*, Paris, 50.

Becker, R., Goldberger, I., Paciuk, M. (2007). Improving energy performance of school buildings while ensuring indoor air quality ventilation. *Building and Environment*, 42, 16.

Cantin, R., Moujalled, B., Guarracino, G. (2005). Complexité du confort thermique dans les bâtiments. *6th Congrès Européen de Science des Systèmes*, Paris, 10.

Déoux, S. and Girard, V. (2010). Bâtir pour la santé des enfants. *MediecoEditions*, Andorra la Vella, Andorra.

Garde, F., David, M., Ottenwelter, É., Ferjani, N., Pothin, É., Lenoir, A. (2010). Outil bâtiment PERENE. *Ile de la Réunion*, 119.

Heschone Mahone Group (1999). Daylighting in schools: An investigation into the relationship between daylighting and human performances. *The Pacific Gas and Electric Company on behalf of the California Board for Energy Efficiency Third Party Program*, 31.

Jindal, A. (2018). Thermal comfort study in naturally ventilated school classrooms in composite climate of India. *Building and Environment*, 142, 34–46.

Kemajou, A., Tseuyep, A., Egbewatt, N.E. (2012). Le confort thermique en climat tropical humide vers un réaménagement des normes ergonomiques. *Revue des énergies renouvelables*, 15, 12.

Kim, J. and de Dear, R. (2018). Thermal comfort expectations and adaptive behavioral characteristics of primary and secondary school students. *Building and Environment*, 127, 13–22.

Kumar, S., Singh, M.K., Mathur, A., Mathur, J., Mathur, S. (2018). Evaluation of comfort preferences and insights into behavioral adaptation of students in naturally ventilated classrooms in a tropical country, India. *Building and Environment*, 143, 532–547.

Lenoir, A. (2013). On comfort in tropical climates, the design and operation of net zero energy buildings. Mechanics and environment. PhD thesis, Université de la Réunion, Saint-Denis.

Malet, J.C. (2017). Impact des différents paramètres de la façade hybride sur le microclimat urbain en milieu tropical humide, à La Réunion. Master's thesis, Ecole Nationale Supérieure d'Architecture de Montpellier, Montpellier.

Pilcher, J.J., Nadler, É., Busch, C. (2002). Effects of hot and cold temperature exposure on performance: A meta-analytic review. *Ergonomics*, 45, 16.

Rodriguez, C.M., Coronado, M.C., Medina, J.M. (2019). Classroom-comfort-data: A method to collect comprehensive information on thermal comfort in school classrooms. *MethodsX*, 6, 2698–2719.

Saddok, A. (2016). Étude du confort thermique des salles de cours des établissements scolaires à différentes typologies. Master's thesis, Université Mouloud Mammeri de Tizi Ouzou, Tizi Ouzou.

Vincent, B. and Gissinger, V. (2016). Les effets du bruit sur la santé. *Acoucité-Observatoire de l'environnement sonore*, Lyon, 30.

7

Urban Green-Blue Grids as a Potential Factor for Regulating Urban Temperature: The Marseille Canal – An Essential Local Opportunity

The city of Marseille is particularly exposed to the urban heat island effect on its abandoned industrial and current commercial sites. While the city continues to develop on its outskirts, which are still largely green, what are the environmental solutions when framing urban transformation, and especially soil sealing? This chapter intends to analyze and juxtapose two French public policies designed to promote a sustainable city. The fight against the urban heat island effect and the introduction of urban green-blue grids in the city could favor the development of new urban planning strategies. In these circumstances, how could the "Marseille Canal", a 19th century infrastructure enveloping the whole city, near to its natural protected spaces, and which is still used as a drinking water supply, become a thick and cooling green path for its inhabitants?

7.1. Introduction

In view of global warming and generalized urbanization, urban overheating – or the urban heat island (UHI) effect, once it has been located – constitutes a phenomenon of notorious discomfort, even a danger for the health of certain populations during the summer season. Indeed, in periods of high heat, the use of outdoor spaces, as well as the use of non-air-conditioned buildings, becomes

---

Chapter written by Séverine STEENHUYSE.

For a color version of all the figures in this chapter, see www.iste.co.uk/aithaddou/transition.zip.

uncomfortable for city dwellers. In addition, the lack of nighttime cooling in the city generates excess electricity consumption to regulate the temperature of homes and workspaces. The urban heat island effect, now widely studied (ADEME 2012, 2017), is still nonetheless difficult to control. It manifests itself as the concordance between multiple factors which can be found in many urban situations (Agam 2018).

Adopted on June 24, 2011, the Regional Climate Plan for adaptation to climate change defined priority fields of action, not only regarding the fight against urban heat islands, but also around water management and health risks and biodiversity. Its adoption by local actors triggered the production of several studies developing the adaptation options available to metropolises in response to climate change[1], and providing benchmarks as well as mitigation principles to address this problem. It was observed that the presence of vegetation and humid atmospheres largely contributed to reducing urban heat.

In another vein, the city also now aims to integrate the promotion of biodiversity (National Strategy for Biodiversity and the biodiversity law of 2016). The question comes down to preserving, promoting or restoring the natural spaces in the city by relying on existing infrastructures, and green spaces in particular, as well as on their connection. The implementation of *urban green-blue grids* develops intervention strategies over the existing space, a process which involves having access to planning and operational documents.

From an interdisciplinary perspective, this chapter would like to focus on the possible intersections between the approaches to regulating urban heat phenomena and those relating to the preservation of urban biodiversity through the development of a revegetation strategy in an urban context. Following what terms, in what urban contexts, based on what incentives and what development policies could these approaches actually overlap and reinforce one another? The point here is to lay the foundations for a possible transversal reflection between two approaches to designing a city which can be more sustainable and better to live in for everyone. What urban policy can be imagined, in particular for the Marseille metropolis, to achieve this cross implementation?

To answer these questions, we will proceed in three stages, starting by recalling the study framework for urban heat island phenomena, in order to establish an inventory of existing mitigation solutions. Then, we will present the approach attached to the urban green-blue grids, and more particularly to the Urban Green Network (UGN) as part of their implementation. Finally, in the context of a local application, an inventory of the existing procedures and situations will be drawn up,

1 Especially in Paris.

making it possible to imagine a concrete policy for the revegetation of the city of Marseille.

7.2. The diagnosis of urban overheating

7.2.1. *Characterization of urban heat islands (UHI)*

"Urban overheating" is a problem that is expressed day and night through a feeling of discomfort. It manifests more markedly in a mineral environment which is not ventilated, or which does not benefit from any cooling phenomenon. Four main factors determine the urban overheating gradient, which can reach between 4°C and 12°C at the edge of a rural area: the urban shape (morphological parameters), the characteristics of coatings, the amount of vegetation (surface parameters) and the concentration of human activities (anthropogenic parameters).

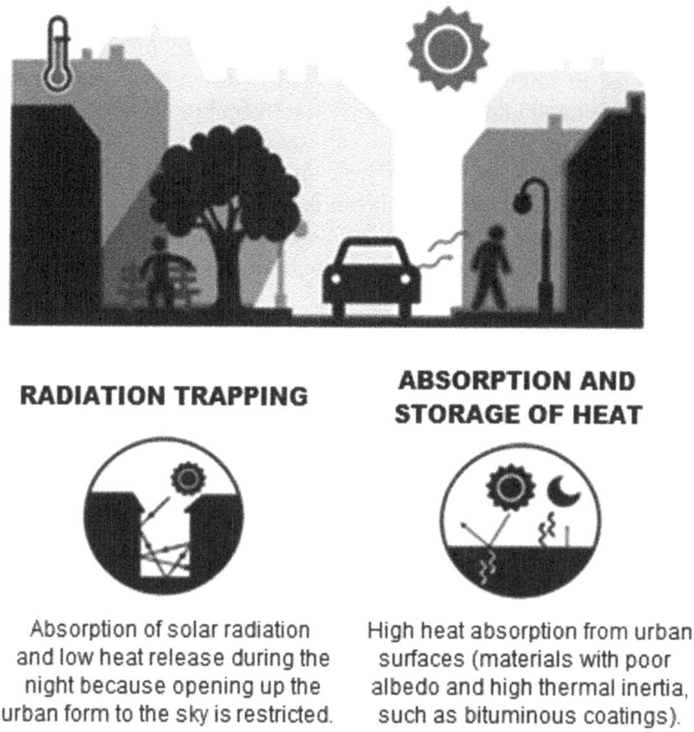

Figure 7.1. *Causes of urban overheating and mitigation through vegetation. Source: (ADEME 2017)*

In order to assess the role of each factor, suitable analytical methods and instruments have been developed to measure the surface temperature of objects (satellite imagery and thermal cameras), as well as air temperature measurements (from two points, a point network, or mobile measurements). We have also integrated the consideration of climatic factors, such as the speed and humidity of the ambient air, relief and orientation, in order to obtain a nuanced analysis of the different situations.

7.2.2. *Urban temperature variation factors*

Thanks to this analysis, it was possible to identify the dominant factors influencing variations in urban temperature, and to anticipate their integration into the design of new urban projects[2]. From this perspective, the presence of water is also an important factor for decreasing urban temperature, even if this benefit remains strictly localized to the water perimeter.

The aquatic element can then be involved in different forms in the urban space: through canals or urban rivers, thanks to the presence of fountains and other associated basins, in rainwater flows, and finally, through sprinkling and misting water systems integrated into the development of recent public spaces[3].

Vegetation also plays an essential role in mitigating urban heat, through the implementation of shading trees which can reduce the effects of sunshine by up to 80%. It also allows the production of associated cool zones through foliage evapotranspiration, which refreshes the air, and the moisture stored in the bedrock, which maintains the humidity.

---

2 One thinks of the relatively "loose" morphology of recent districts such as Confluence, Lyon, available at: http://www.lyonconfluence.fr/ressources/document/191105dpgeneriquelyon confluence.pdf, p. 13.

3 The water mirror of Bordeaux was created by the landscape designer Michel Corajoud and fountain designer Jean-Max Llorca in July 2006. It offers the largest reflecting pool in the world (3,450 m^2), facing the Place de la Bourse in Bordeaux, between the Quai de la Douane and Quai Louis 18. It operates every day from 10 a.m. to 10 p.m., following a specific cycle: 3 minutes for filling, 15 minutes of mirror effect, 5 minutes for emptying and 3 minutes of fog. Its service is interrupted in the winter season for maintenance; available at: https://www.bordeaux.fr/l10812/miroir-d-eau.

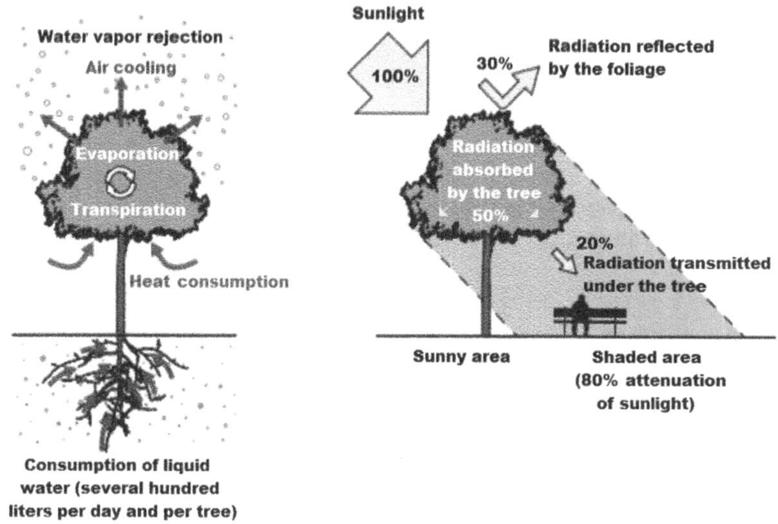

Figure 7.2. *How vegetation reduces urban overheating. Source: (ADEME 2017)*

In addition, the existence of urban parks has a notable effect on lowering the temperature within them, as well as on their surroundings, as shown in the document of night temperature readings near the Buttes Chaumont Park, carried out in Paris in 2011 as part of the Climate Plan.

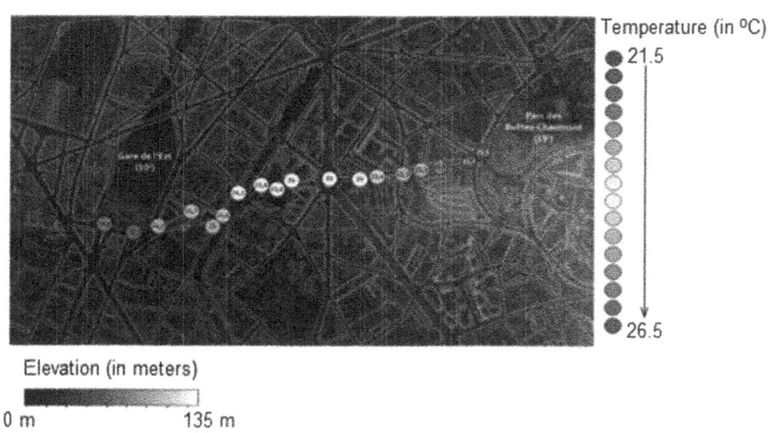

Figure 7.3. *Air temperature readings recorded between Gare de l'Est and Buttes Chaumont Park, Paris, on August 1, 2011 at around 10 pm (8 pm UTC). Source: (ADEME 2012)*

More detailed analyses also make it possible to identify the differences in temperature behavior depending on the types of vegetation (trees, beds or lawns), and the ground coverings used for traffic (earth, stabilized, soil/grid mixture, etc.).

In addition, urban forms and ground cover materials generate configurations in terms of heat recovery and night ventilation that are of varying degrees of benefit. In particular, earthy areas always remain cooler than artificial coverings, and the darker the covering, the more it stores and releases solar heat.

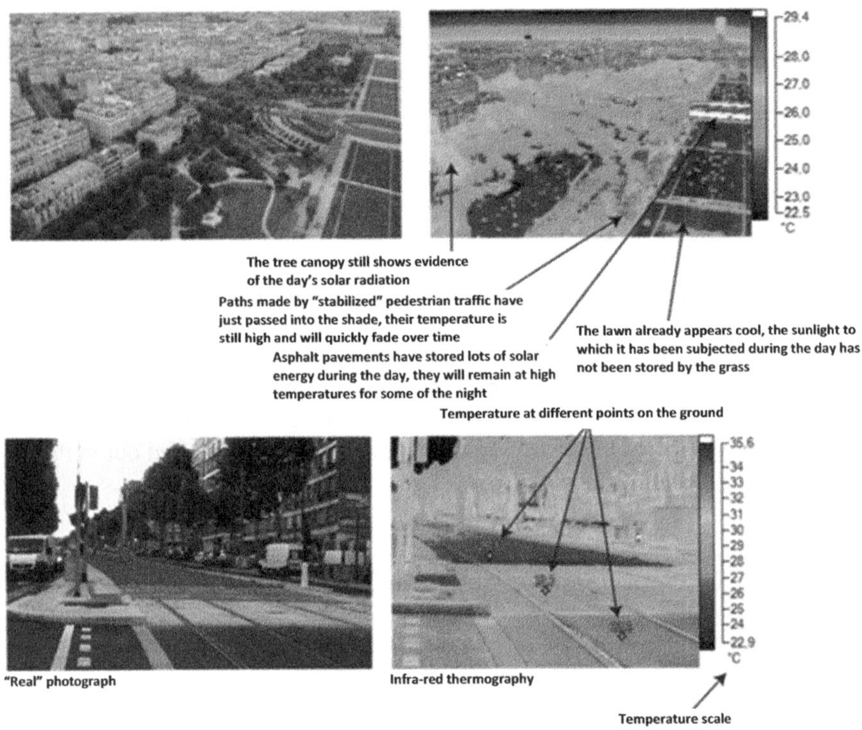

Figure 7.4. *Aerial views of Champs-de-Mars taken on August 2, 2011 at around 8 pm (6 pm UTC), just after sunset. Source: (ADEME 2012)*

Finally, in general, the analysis of Parisian urban forms makes it possible to analyze the different behaviors of public spaces and built typologies.

Areas of "relative coolness" are particularly identified in the shade of Haussmann buildings, and we see that the reflection of urban heat is significant in large unshaded public spaces (like the esplanade of the George Pompidou center). Finally, roofs tend to release more heat than the ground does.

Sunlight in two public spaces is very different on August 1. At the end of the day, a difference of 2°C is observed between the narrow *Rue Geoffroy l'Angevin* and the *Esplanade de Pompidou*. During the night, the temperature of the esplanade will likely drop more quickly due to a clear sky

Figure 7.5. *Calculation of the insolation received on August 1, 2011. Source: (ADEME 2012)*

Regarding urban forms and their architectural design, it is also necessary to mention the possibility of greening the roofs and facades, the effects of which are certainly more subtle, but, all the same, enable an alleviation of heat reflection. Indeed, the greening of roofs and facades (climbing plants or living walls) reduces the amount of heat transferred to the building, thanks to evapotranspiration and the shading obtained. In addition, surface humidity helps to cool the outdoor ambient air.

Since these treatments depend on specific technical procedures, they still generate additional construction costs. In addition, the revegetation of spaces is currently not feasible in all circumstances (heritage context, social acceptability).

Green roof in Parisian region *Living wall (Gare de l'Est)*

Figure 7.6. *Green rooftops and living walls in the Parisian region*

Finally, concerning the acknowledgment of factors which contribute to warming, it is important to add that commuting or functional movements associated with housing or activities constitutes a significant factor in urban heat increase.

All in all, the factors contributing to the rise in urban temperature are multiple and of various origins, and are also cumulative. Because of this, there are essentially two types of situations to consider for intervention in the direction of urban cooling: virgin land, which offers great freedom for design, and the rehabilitation of existing neighborhoods, identified as a priority. In both cases, taking urban cooling into account leads to an additional cost in the production of the built tissue, as well as additional maintenance, which will have to be faced[4].

4 The exception that confirms the rule, available at: https://villedurable.org/2017/05/05/eco-quartier-sans-surcout-la-demonstration-de-palaiseau/.

7.2.3. *Remedy to UHI effect: the essential role of revegetation*

The UHI effect is therefore enhanced by different factors which make up the very characteristics of urban space: urban forms, built density, human activities, materials of a mineral nature, the absence of vegetation, the presence/absence of water, soil artificialization, the treatment of roofs and facades, protection against prevailing winds, orientation, etc. One of the main causes for heat islands is due to the urbanization and development model of the dense city, which by nature concentrates many human activities that emit heat in small and protected areas.

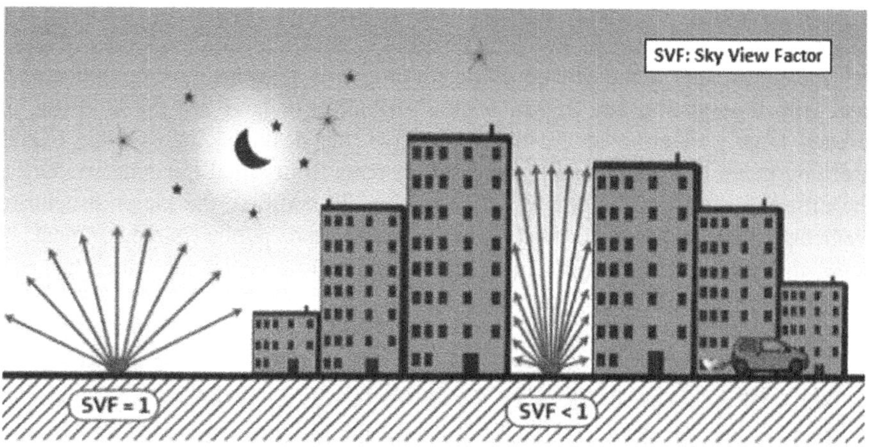

Figure 7.7. *The sky view factor (source: APUR)*

Nevertheless, as the analyses carried out in 2007 within the framework of the Paris Climate Plan show, there are simple and effective mitigating factors. The reintroduction of natural spaces, and more generally, plants to the city, are effective solutions to fight against the UHI effect, concludes the Ademe report (Ademe 2012). For example, wooded areas are 2 to 8°C cooler than the rest of the city.

Vegetation acts in several ways to regulate urban temperature. On the one hand, it makes it possible to stabilize the air temperature by retaining water in the leaves, thanks to evapotranspiration. The increase in ambient humidity also causes a drop in air temperature. Planting individual or rows of trees, increasing the number of planted spaces, urban parks and natural areas are therefore essential to reduce the intensity of heat islands, by providing public spaces with shade and cool[5]. Finally, the revegetation of horizontal above-ground (roof terraces) or vertical (facades) surfaces also makes it possible to lower the temperature to a certain extent.

All of these parameters and their accumulation then actively participate in cooling the urban space.

7.2.4. *Conclusion: simple solutions... but difficult to implement*

The greening of urban spaces is a determining factor in lowering the temperature. While it seems perfectly possible to ensure a certain quantity of green or vegetated spaces in the new development programs, it seems more complicated to deal with the existing city in which the distribution of land and the land use are already largely determined. In this perspective, the desire to increase the overall greening index of a municipality necessarily stems from a real greening strategy at the different scales of the territory and the city. In addition, the major structuring elements, such as wooded areas, urban parks and public gardens, must be supplemented by the development of diffuse and private vegetation which makes it possible to perpetuate the cooling effect felt.

The benefits created by the presence of vegetation in urban public spaces therefore encourage us to look for intersections with other policies which would enhance the role of plants in the city. The policy of urban green-blue grids is also based on an enhancement of vegetation expansion... To what extent could they contribute to cooling the city?

7.3. Urban green-blue grids to limit urban overheating?

Following the assessment of the various solutions identified to remedy urban overheating, we are tempted to question another type of public policy which is also related to the enhancement of urban vegetation, but this time from the point of view of biodiversity.

---

5 Among the main species planted on public roads, the plane tree, the hackberry and the lime tree are excellent shade trees and provide comfort for pedestrians in public spaces. The hackberry is a deciduous tree that grows well along avenues and provides excellent shade. The lime tree is a deciduous tree with a short trunk and rounded crown and provides excellent shade in summer. The resistant plane tree provides very generous shade, it has rapid growth, long life and is resistant to pollution. It grows well along avenues or in shady areas (ADEME 2012).

Since the *Grenelle de l'Environnement* (2007), in the context of the desire to reduce greenhouse gases (GHGs) and to preserve biodiversity, a more sustainable way of considering the city was proposed. In particular, it relied on the implementation of urban green-blue grids, whose primary concern was to maintain ecological continuities in order to preserve ecosystems and species habitats in the context of a decrease in biodiversity.

7.3.1. *The urban green-blue grid: definition*

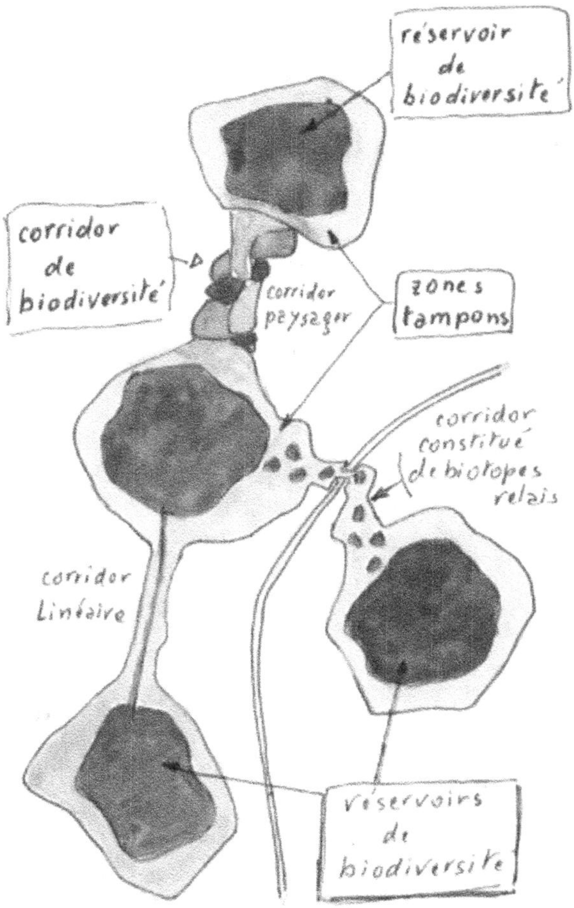

Figure 7.8. *Operating diagram of an urban green-blue grid (source: Sauvegarde de l'Anjou, credits: Alain Lasserre)*

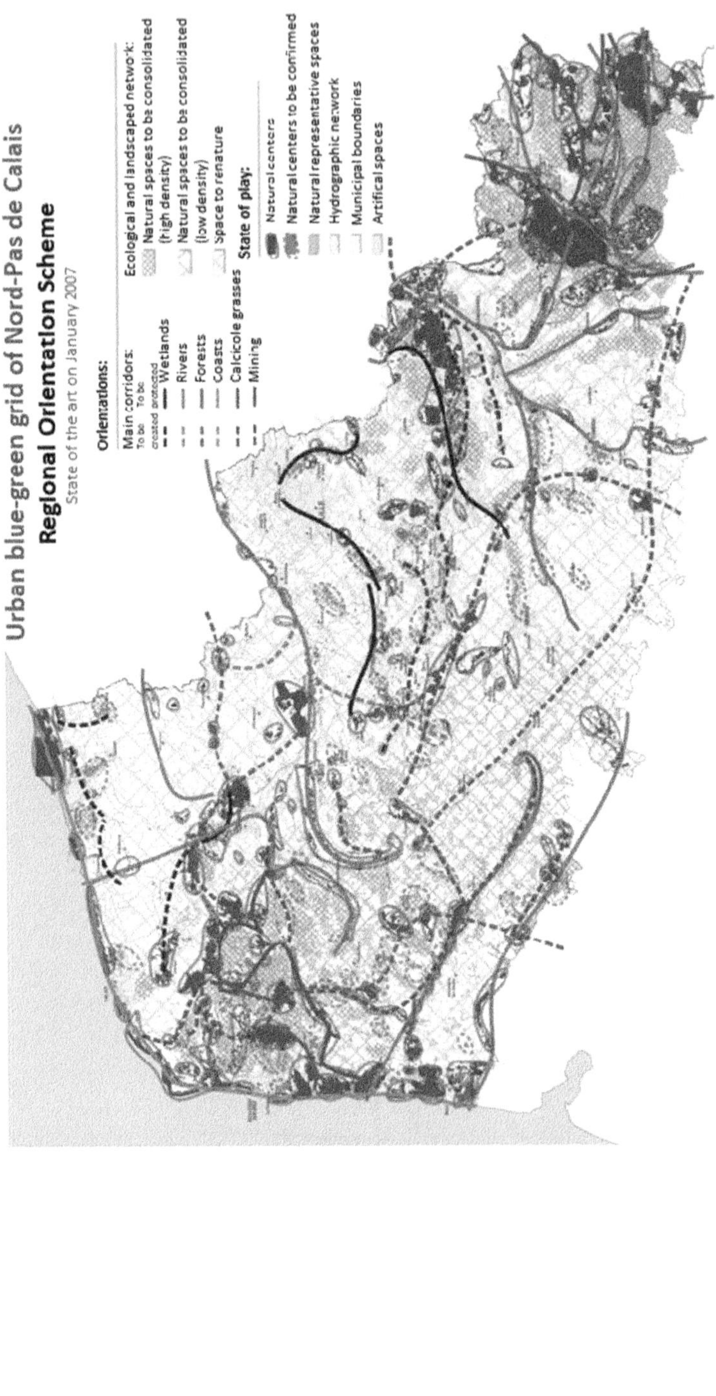

Figure 7.9. *Territorial scale of the urban green-blue grid implementation at Nord-Pas de Calais (source: http://dismagazine.com/dystopia/67488/the-new-metropolitics-of-nature-framing-resilient-urban-ecologies/)*

Urban green-blue grids aim to identify the living spaces of species and their travel routes in order to preserve them in good condition. It is defined on the basis of the analysis of existing biodiversity, detecting biodiversity "reservoirs" and "corridors" which define the ecological fabric of a territory (see diagram).

Defined by article 121 of the Grenelle 2 Law, it has been transcribed into Article L371-1[6] of the Environmental Code. Its goal is to prevent the fragmentation of terrestrial and aquatic "ecological environments". Its implementation is based on the implementation of the *Schéma Régional de Cohérence Ecologique* (SRCE, Regional Scheme for Ecological Coherence) which involves concerted work involving the state, communities and local actors to produce a state of knowledge about the local environment. Thus, in each region, between 2011 and 2015, SRCE sessions were carried out in order to be translated into regional and local strategies, rules and concrete actions, in SCoTs (*Schémas de Cohérence Territoriale*, Territorial Coherence Schemes) and urban planning documents.

Today, the transition to the operational phase of these development documents requires the increased involvement of local partners as well as the synergy of existing tools because the issues identified in the SRCEs are numerous: restoring degraded environments, improving the transparency of ecological transport infrastructure, establishing new agro-ecological infrastructures, and supporting innovative projects in favor of the restoration of the urban green-blue grid. To meet these challenges of ecological functionality, several planning and management tools are available, such as: contractualization, regulatory protection measures, planning and urban planning tools, land actions, or technical and financial support systems. The French Agency for Biodiversity[7] is responsible for supporting the various actors involved in biodiversity and planning.

7.3.2. *The UGN and its implementation*

In urban areas, the identification of habitat areas and species journey trajectories is much more complex and reduced. The urban green-blue grid is particularly

---

6 Article L371-1. I. The objective of the green and blue grids is to halt the loss of biodiversity by participating in the preservation, management and restoration of the environments needed for ecological continuity, while taking into account the human activities (particularly agricultural ones), in rural areas. This legislative article of the environment code also aims "to improve the quality and diversity of landscapes".

7 Grouped since January 1st, 2020 with the National Hunting and Wildlife Office within the French Office for Biodiversity (FOB).

limited due to soil fragmentation, the obstacles built, real-estate pressure and anthropogenic uses. In the city, "biodiversity reservoirs" are assimilated to parks, forests and wetlands, while "corridors" correspond to abandoned areas, cycling paths, small gardens, squares, cemeteries and stadiums (Kervadec 2012).

In this context, the implementation of the UGN cannot only focus on the preservation of environments. It requires the deployment of a specific approach to articulate the connection between ecological questions and existing uses, or to promote actions which encourage the presence of nature in the city (Clergeau 2016).

The initial ecological quality of an urban green space or of an environment can then be assessed through the various tools employed by local managers. The Biotope Area Factor (BAF)[8] makes it possible to assess the environmental quality of a plot, an island, a district or a larger territory, for example.

Ecological diagnostic methods can also be developed by internal departments, such as the Green Spaces Department of the city of Lyon (Lyon 2014)[9], or by ecologists, more generally[10].

The relevance of the implementation of the green network in an urban environment is, in all cases, based on the ecological diagnosis of the environments, which can be included in a conservation-restoration or valuation project.

---

8 A biotope coefficient or biotope coefficient per area (BCA) or biotope coefficient per plot designates the part (the percentage) of a developed area that will be permanently devoted to nature (vegetated area and/or favorable to local ecosystems and local species) in the total area of a plot that is or is to be developed. This coefficient aims not only to conserve a certain quantity of non-waterproofed and non-artificialized land, as well as its ecosystemic services. It applies both to plots intended for new construction and built areas, undergoing renovation as "space consumption management tools". It was formalized by the ALUR Law (March 2014) which proposes it without imposing it on SCoTs and PLUs which, on the other hand, must take biodiversity into account. Available at: https://www.o2d-environnement.com/wp-content/uploads/2017/12/coefficient-de-biotope-ademe.pd.

9 Tatiana Bouvin, Green Spaces Department, Head of Sustainable Development, City of Lyon. Available at: http://www.capitale-biodiversite.fr/experiences/diagnostic-de-la-valeur-ecologique-des-espaces-verts-lyonnais.

10 Available at: https://www.nature-en-ville.com/sites/nature-en-ville/files/2018-03/fiches-Natureparif2-fiche3.pdf.

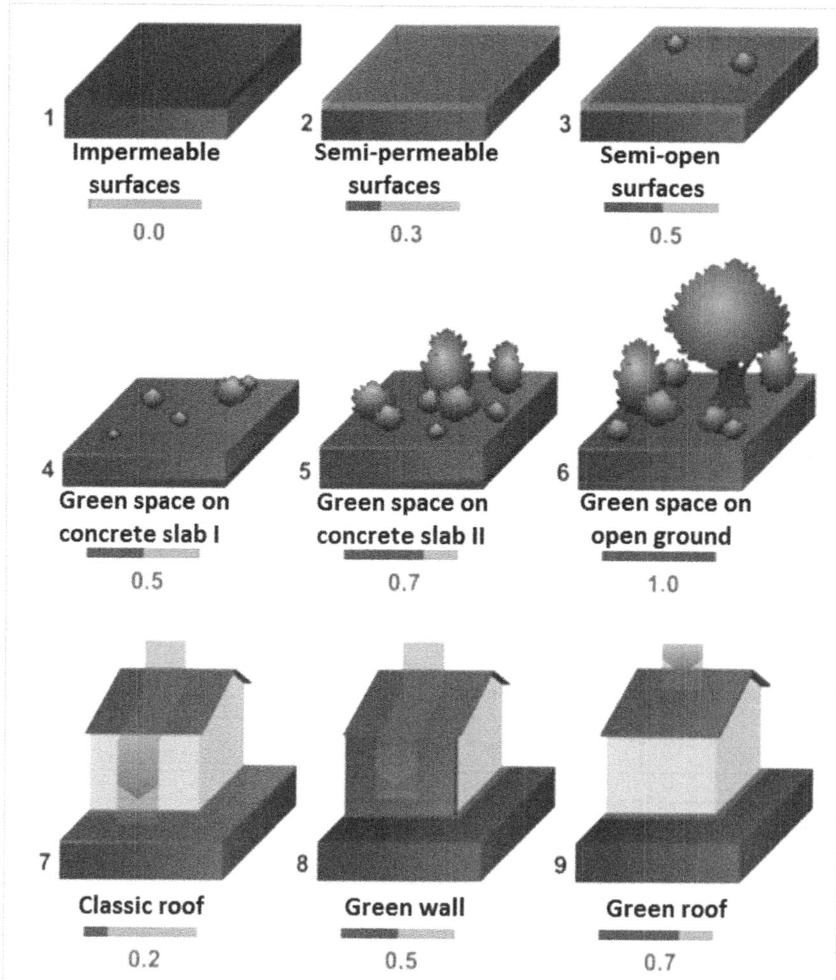

CBS: The eco-friendly area / plot area

The eco-friendly area is calculated from the different types of area that make up the plot

Eco-friendly surface: (surface of type A x coef. A) + (surface of type B x coef. B) + ... + (surface of type N x coef. N)

Figure 7.10. *Different city surfaces possible and how eco-friendly they are. Source: (ADEME 2017)*

Criteria assessed during this diagnosis

Site identity: Site protection on urban planning documents (classified wooded areas, etc.), previous pollution of the site.

Recovery rate: Land use (mineral/water/living surfaces), covering layers of different vegetation.

Threats: Frequentation of the site, aberrant uses, invasive species, lighting, permeability of the site to species.

Flora: Number of species, diversity of flowers, mixture of woody species, proportion of site to species.

Fauna: Emblematic species, micro-habitats, devices installed by humans, presence of birds, soil life.

Soil: planting space, soil depth, ventilation and compactness, stones/pebbles, mulching.

Rate of cover

1	Soil permeability, in % relative to the total area of the site	Impermeable mineral surface	a	O 0% O < 15% O 15% – 49% O 50% – 85% O > 85%
		Permeable mineral surface	b	O 0% O < 15% O 15% – 49% O 50% – 85% O > 85%
		Water surface	c	O 0% O < 15% O 15% – 49% O 50% – 85% O > 85%
		Non-mineral permeable surface	d	O 0% O < 15% O 15% – 49% O 50% – 85% O > 85%
2	Covering of strata, in % with respect to the non-mineral permeable surface defined above	Wooded (more than 7 m)	a	O 0% O < 15% O 15% – 49% O 50% – 85% O > 85%
		Arborescent (between 3 and 7 m)	b	O 0% O < 15% O 15% – 49% O 50% – 85% O > 85%
		Shrubs (less than 3 m)	c	O 0% O < 15% O 15% – 49% O 50% – 85% O > 85%
		Herbaceous	d	O 0% O < 15% O 15% – 49% O 50% – 85% O > 85%
		Mossy	e	O 0% O < 15% O 15% – 49% O 50% – 85% O > 85%

Site identity

1	Identification	EI	a	
		Site name	b	
		Address	c	
		Typology	d	O - A - O - B - O - C - O - D - O - E - O - F -
2	Area			O - A - O - B - O - C - O - D - O - E -
3	Insertion of the site within the urban green-blue grid			O - A - O - B - O - C - O - D -
4	Ecological continuation	Distance to a space of the same type	a	O < 20 m O 20 – 100 m O 100 – 500 m O > 500 m
		Minimum distance to another natural area	b	O < 20 m O 20 – 100 m O 100 – 500 m O > 500 m
5	Site sustainability			O - A - O - B - O - C -
6	Site history			O - A - O - B - O - C -

Table 7.1. *Source: (Lyon City 2014)*

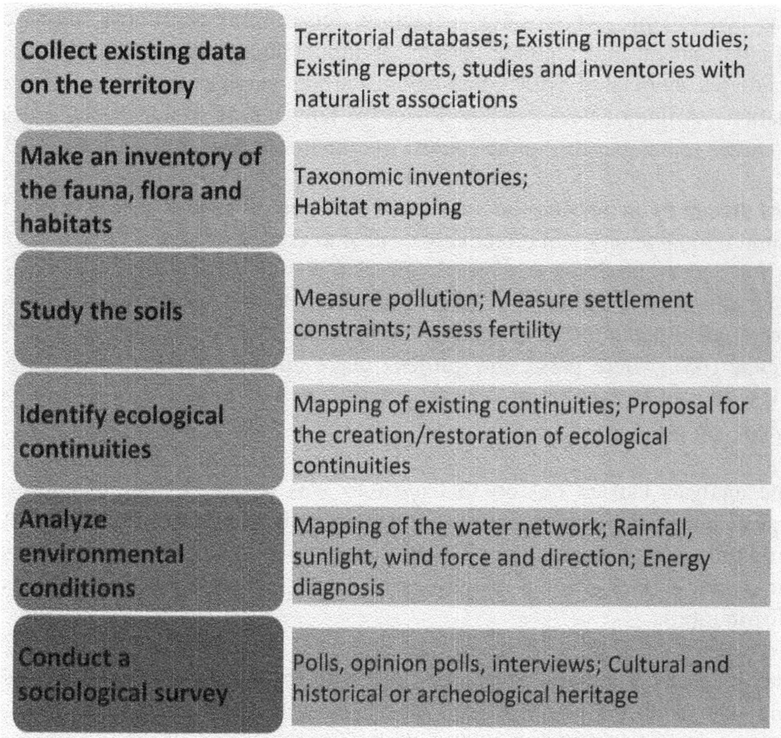

Figure 7.11. *Elements to be engaged for carrying out an "ecological diagnosis" (source: Naturparif 2018)*

For Nathalie Blanc (2012), following the ANR (*Agence National de la Recherche*, National Research Agency) program on urban green-blue grids (2009), "it is a question of mobilizing heterogeneous knowledge about the environment and of turning it into a collective research and space management issue". On the urban Plaine Commune (93), Philippe Clergeau immediately addressed the UGN question using the notion of "ecosystem service" as a departure point (Clergeau 2016). Ecosystem services have been defined as the benefits that humans can draw from nature through three categories: supply (supply of food, materials, etc.), regulation (air purification, pollution regulation, etc.) and cultural (aesthetics, education, atmosphere, etc.) (Millennium Ecosystem Assessment 2015).

In the city, parks, public gardens, wastelands, road borders, private gardens, etc. constitute natural spaces (ECN, *Espaces à caractère naturel*) (Clergeau 2007) which provide ecosystemic services for recreation, leisure and education, improve the living environment and regulate certain environmental problems (fixation of dust and particles associated with combustion, carbon storage, rainwater infiltration, pollination). A line of trees can also lower the temperature of a street by 3 degrees and we must remember this in the context of climate change (Bowler *et al.* 2010).

It is indeed by increasing the number of high-quality natural spaces (ECNs) that a minimum biodiversity will effectively be able to establish itself.

How to associate human activities within these natural spaces without hindering their ecological potential? What ecological network could be possible on the Plaine Commune (93), where parks and gardens make up the dominant share of ECNs (65%), with a small number of very large parks, and with numerous infrastructures (railways, highways)?

The analysis carried out on this territory was based on four diagnoses: the ecological diagnosis, carried out at 3 different scales (regional, district, plot) and identified the components of the UGN following three categories of primary nuclei, secondary nuclei and potential relay spaces (analysis of their areas, shapes, distances and distribution).

The landscape approach was carried out with regard to the ECN at two scales, making it possible to identify large landscape structures, on the one hand, and singular or recurring motifs that form the basis of intercommunal identity, on the other.

The urban diagnosis compiled an inventory of existing projects which concern 50% of the territory's surface, and constitute both a risk and an asset for the UGN by offering the possibility of reinventing ecological continuities, but also entailing the risk of eliminating the ecological animal and plant wealth of "hollow teeth", or wasteland spaces.

Finally, concerning the diagnosis of practices, the concept of "sociotope", or of nature as a support for diverse uses[11], made it possible to combine the planning of public spaces with their ecological preservation. People were questioned in order to identify their "natural territories", that is to say, their urban green space practices as

11 If one goes to a particular garden or public park, it is thanks to its accessibility, and also because one finds the amenities suitable for the activity ones wishes to carry out there, because one feels good there, one finds one or more interesting experiences and one has a particular mental representation of it, a kind of sensitive caricature of the place.

territorialized from their home[12]. Five sociotopes have emerged from the Plaine Commune, thus defining "living spaces" (equipment), "connection spaces" (avoidance path), "dependency spaces" (annex to another urban function), "reserves" (little frequented but appreciated) or "island spaces" (isolated).

The mapping of sociotopes and their respective areas of attractiveness has led to the localization and quantification of gaps in public green spaces, as well as to the identification of the nature of these gaps.

Figure 7.12. *Distribution of ECNS classified according to five types of sociotopes and their areas of influence. Source: (Linglart 2016)*

Synthesis maps produced an ideal potential network combining the territory's current situation with the feasibility of transformation projects.

12 The natural territories that emerge from the survey are sometimes limited to a few hundred meters around the place of residence or they may extend over ten kilometers, outside the limits of intermunicipal cooperation. These territories are made up of public green spaces (squares, gardens, squares and parks), but also wasteland (vacant spaces awaiting development and occupied by pioneer vegetation), pedestrian paths, family and collective gardens, etc.

During the working sessions with the elected representatives of the territory, the ECNs also appeared as elements the generate a real territory quality and improving the living environment, the health of inhabitants, the quality of the ambient air, and offering spaces for various recreational purposes which have been integrated in the urban planning documents of the territory (SCoT and PLU), as well as in the territory's community organization guides.

7.3.3. *Link with the "nature in the city" challenges*

The implementation of the UGN here meets the approach for valuing nature in the city resulting from the plan "Restoring and enhancing nature in the city"[13].

Therein nature was defined as everything that is alive in an urban environment, as opposed to the city's mineral components. Green spaces are considered through a range of seven types of spaces which can be implemented for the city's greening project: large parks, squares, collective gardens, green corridors, traffic lane surrounding areas, building yards and buildings (CEREMA 2015). What associated planning instruments should be conceived and how to consider the distribution and continuity of these spaces?

7.3.4. *The implementation of the urban green-blue grid*

In this perspective, the implementation of the urban green-blue grid is based on five types of action: protection[14], the development of knowledge[15], contractualization[16], planning[17] and real-estate control[18].

---

13 On November 12, 2010, translating the *Grenelle de l'Environnement* 76 commitment, the State Secretary for Ecology presented the main lines of the multi-year action plan (2010–2012) intended to "restore and enhance nature in the city", as well as its multiple functions. This plan was expected to help "anchor the city within its geography and its natural environment", "to develop natural spaces in quantity and quality" and "to promote a culture and a shared governance of nature in the city". Among the 37 actions making it up, the plan notably provided for the creation of atlases of biodiversity in large urban areas by 2012, as well as the development of urban green-blue grids. Available at: https://www.banquedesterri toires. fr/un-plan-dactions-pour-restaurer-la-nature-en-ville.

14 This includes National Parks (NP), Hunting and Wildlife Reserves (HWR), Nature Reserves (NR), Prefectural Biotope Protection Decrees (PBPD), Priority Biodiversity Zones (PBD), Integral or Managed Biological Reserves (IBR/MBR), fishing reserves, Regional Natural Parks (RNP), sites of the Coastal Conservatory, classified sites, the Natura 2000 network, the *Grands Sites de France*, classified water sources, and Wetlands of Special Environmental Interest (WSEI).

Most importantly, success in the implementation of this policy lies in the consultation of actors upstream of the projects, and with the ability to mobilize funding in order to implement acquisition and/or management policies.

7.3.5. *Conclusion: heat island and the urban green network: possible overlaps?*

The UGN problem involves the analysis and monitoring of ecological environments identified in different contexts and places environmental issues at the heart of development dynamics. However, the procedures and requirements are sometimes difficult to appropriate for local actors. The requirement to adapt the scales of intervention from the region to the local level makes their implementation particularly delicate within very heterogeneous territories. The density of buildings, the number of actors and the difficulty of defining biodiversity in urban areas scientifically further increase the obstacles to implementing UGN (Clergeau 2016). To what extent can its realization contribute to cooling urban public spaces?

At this stage, two very distinct types of situations appear: intervention on existing tissues and the creation of new sectors. In a new district, planting strategies, the degree of waterproofing and land density are still parameters which can be controlled more easily than on already established tissues. In the case of existing tissues, the revegetation of the urban space is more a matter of a territorial revegetation strategy based on the designation of vegetated travel axes (malls, avenues, railway lines, alignment plantations, etc.), the definition of public plots to

---

15 This includes the National Inventory for Natural Heritage (NINH), the Municipal Biodiversity Atlas (MBA), the Landscape Atlas, the National Biodiversity Observatory (NBO), the CARMEN site, and the different Geographic Information Systems (GIS).

16 Based on agri-environmental measures, management or provision agreements, Rural leases with Environmental Clauses (REC), Park Charters, etc.

17 In the SCoT and the PLUi, through the PADD, the Orientation and Objectives Document (OBD), the Landscape Quality Objectives (LQO), and the specific zoning of the PLUs (zones A and N, Ecological Continuity Areas, Classified Wooded Areas (CWA), article L123-1-5 of the urban planning code, the Development and programming Guidelines (DPG), the Biotope Coefficient, the Perimeters for the protection and enhancement of peri-urban agricultural and natural areas (PUANA).

18 Through purchase, the Right to Urban Preemption (RUP), expropriation, PAEN, the Protected Agricultural Zone (PAZ), the Conservatory for Coastal Space and Lake Shores (CCSLS), the Conservatory for Natural Areas (CNA), the Departmental Council for Sensitive Natural Areas (DCSNA), Land Development and Rural Establishment Companies (LDREC), Public Land Establishments (EPF), regional agencies.

be greened (green spaces, outdoor spaces of public facilities, etc.), the identification of rules for greening private property (perimeter of buildings, greening the feet of buildings, green facades, etc.), and an urbanization policy that encourages the planting and preservation of peripheral agricultural plots.

These crossed working hypotheses only take on their full meaning through their concrete situation. The analysis of the Marseillaise urban situation will be the chance to formulate the opportunities and limitations of the perspective discussed here.

7.4. The Marseille case

From 2010, the *Projet d'Aménagement et de Développement Durable* (PADD, Local Planning and Sustainable Development Project) identified large peripheral natural spaces (massifs, hills, sea, rivers) and agricultural areas of the city of Marseille as the "support for an urban green-blue grid". In addition, the municipality formulated the ambition to "constitute a regular and structuring plant grid", through the creation of a park network, workers' gardens, heavily vegetated public spaces and so on. However, this ecological framework called the "blue-green grid" was quickly limited to two types of actions: the identification of the rare public ECNs and the engagement of some landscape elements (traffic route plantations, forestation, parks and gardens) identified as *Espaces Boisés Classés* (EBC, Classified Wooded Areas).

By observing the reality of local functioning more closely, it appeared that above all, "between a logic of environmental enhancement and a process of undeniable real-estate consumption", appointed representatives seemed to reinforce what researchers have identified as "green schizophrenia", that is to say, a set of contradictions and ambivalences leading to paradoxical behaviors with regard to nature, and manifested in an opportunistic and unstructured consumption of the natural spaces still available.

7.4.1. *The Marseille metropolis*

The Métropole d'Aix-Marseille Provence (AMP) now includes 92 municipalities over a territory of 3,100 km^2, around the two urban centers of Marseille and Aix-en-Provence. It brings together a largely urban population of 1,850,000 inhabitants, 60% of whom live in Marseille or Aix. Despite the presence of the Mediterranean nearby, the metropolis is largely affected by the UHI phenomenon, which is amplified by the territory's hilly topography.

In Marseille, the mineralization and waterproofing of soils generate high surface temperatures, particularly in the former industrial zones north of the city and east of the port (railway lines, activity zones, mineral public spaces, gullies, etc.).

This specificity of the Marseille area due to its former industrial activity generally raises the question of the reconversion of wasteland, and the formulation of projects capable of reintegrating green spaces in a tense economic context.

Figure 7.13. *Marseille urban heat islands. For color legend see Figure 7.14. Source: (Regards ENVIRONNEMENT 2018)*

The other essential urban issue to be addressed in the city of Marseille is the question of commercial areas which generate real urban heat spots, due to the highly mineral dimension of their developments, the absence of vegetation and shadows, and the dark asphalt coating of the road.

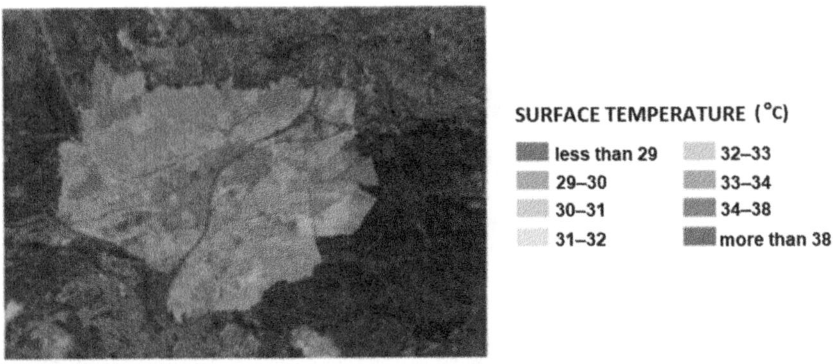

PLAN-DE-CAMPAGNE *The commercial area of Plan-de-Campagne is an island of concentrated high surface temperatures, in the heart of unspoiled nature (agricultural areas and natural areas). The presence of large, but simply built volumes and large parking areas with little vegetation accentuate the surface heat.*

Figure 7.14. *Surface temperature of Marseilles, commercial zone Plan-de-Campagne, August 2015/2016 Source: (Regards ENVIRONNEMENT 2018)*

7.4.2. *An urban context in full reconfiguration*

On the economic level, Marseille is currently in a phase of economic reconversion which, since 1995, has been based on the *Euroméditerranée* operation, an urban renewal operation launched by Robert Vigouroux, mayor of Marseille, and the state, over a perimeter of 310 hectares.

Since 2007, "*Euroméditerranée Act 2*" has extended this perimeter by 170 additional hectares, to now cover an area of 480 hectares in the heart of the city.

This major land conversion operation between the commercial port, the Old Port and the train station has enabled the creation of new urban sectors integrating environmental issues, as well as the programming of positive energy buildings.

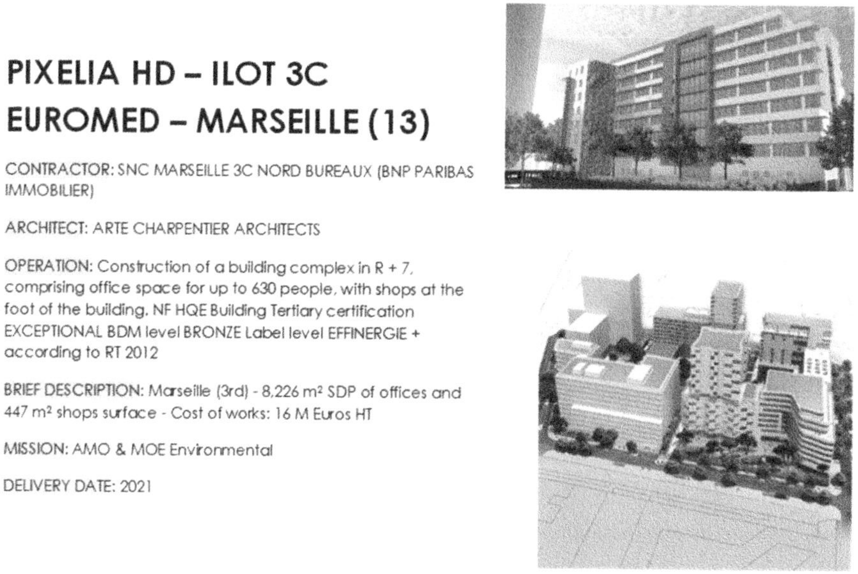

Figure 7.15. *One of the blocks proposed by the Euroméditerranée Act 2 project. Source: https://www.oasiis.fr/reference/pixelia-hd-ilot-3c-euromed-marseille-13*

The Smartseille project, which falls within the scope of the 2007 extension, received the *EcoCité* label in 2009 from the Ministry of Ecology, Sustainable Development and Energy, which promotes "architectural, social and energetic innovation at the city level". As such, the sector was imagined as an eco-district through the construction of a demonstrator island – the Allar island – on a 2.7 hectare plot, accommodating 58,000 m^2 of floor space, 385 positive energy housing units including 100 social housing units, a seniors' residence with 100 beds, 27,500 m^2 of offices, 3,000 m^2 of shops, a hotel with 90 rooms and a nursery with 35 cradles. The Smartseille eco-district now has 4,000 inhabitants and 2,000 jobs, as delivered in 2018.

The *Fabriques* project is supported by two promoters (Linkcity, Bouygues Construction and UrbanEra, Bouygues Immobilier) who have produced a new construction over an area of 14 hectares. The district is designed to be woven with pedestrian paths, shaded alleys and public places, encouraging walking and soft mobility. Around 170,000 m^2 of new housing has been planned, 20,000 m^2 of local shops, schools, training centers, student, senior and hotel residences, a media library and a nursery. Large spaces will be dedicated to companies, with offices and coworking spaces (44,000 m^2).

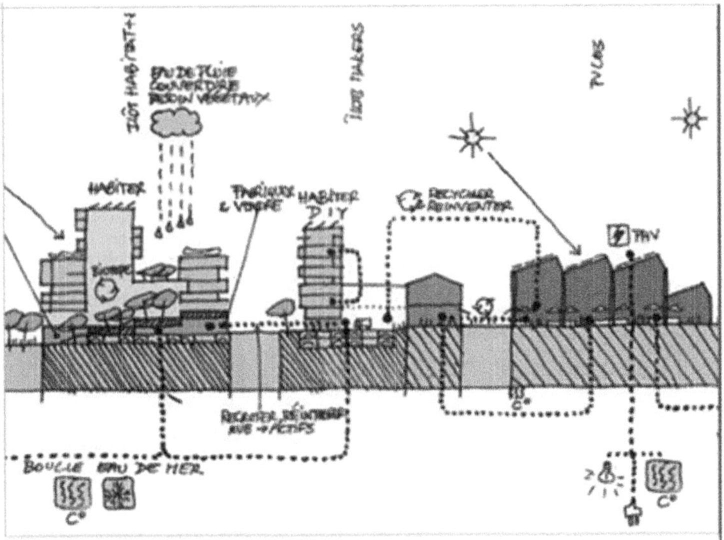

Figure 7.16. *An eco-district model for the Mediterranean. Source: https://www.lesfabriques.fr/decouvrir/*

Figure 7.17. *Parc des Aygalades project plan and details. Source: https://www.euromediterranee.fr/projets/parc-des-aygalades*

The second phase will also include the creation of a large public park around the rehabilitation of the *Ruisseau des Aygalades* (currently still very polluted due to upstream industrial activities). In addition, several other parks and public spaces have been scheduled (Parc Bougainville of 4 ha).

The project is particularly ambitious and exemplary from an environmental point of view. The Mediterranean characteristics of the site (sunshine, views, sea, wind, noise, etc.) have been modeled in order to optimize the management of resources and to provide as much natural comfort as possible – by moving the dwellings away from noisy areas, by avoiding heat islands and wind corridors and promoting outdoor spaces.

A clear mineral treatment will be preferred for the realization of the facades, in order to reflect the light while preserving the freshness of the building. Around 70% of the district's energy will come from renewable energies (sea and sun). The eco-district will benefit from the demanding BiodiverCity label and the housing units will receive environmental certification (HQE, BDM, etc.)[19]. Specific attention to biodiversity has also been given by the INDIGO design office, a Sustainable Development Research Office.

7.4.3. *A rich, natural but ambivalent context*

Legend has it that when the Phocaeans landed on the Provençal coast, they were impressed by the thick forest of oaks, ash, alder and pines which covered most of the territory. It was then necessary to wait until the end of the 15th century and the arrival of King René for attention to be given to urban revegetation through the creation of Marseille's first acclimatization garden. Then, between the 16th and 18th centuries, the development of country houses and their gardens on the immediate city outskirts took place (Borruey 1983). The first urban parks were created in the 19th century with the Longchamp garden (1862) and the Borély park (1863). In 1892, the surface area of Marseille's green spaces was estimated at 56 hectares for a population of 340,000 inhabitants (that is to say, 1.6 m^2/inhabitant), but in 1968, this figure reached 66 hectares for about 888,820 inhabitants (0.7 m^2/inhabitant).

Faced with this alarming observation, Gaston Defferre, mayor of Marseille from 1953 to 1986, decided to set up a vast program to create and maintain public green spaces. From the 1970s, he created the Department of Green Spaces and the Environment (SEVE, *Service des espaces verts et de l'environnement*), which rapidly evolved into a Department of Ecology and Green Spaces (DEEV, *Direction de l'écologie et des espaces verts*), with the purpose of creating 1000 green spaces by the 1990s (Saint Martin 1998). The properties and country houses of the 18th and 19th centuries were then inventoried as part of the preparation of the 1978 Land Occupation Plan (POS, *Plan d'Occupation des Sols*) so that they could be protected. The creation of the beaches was then decided in 1967 by the City Council. After 1975, it was supplemented by the creation of a large strip of greenery of 13 hectares, which nowadays extends over 28 hectares (subway embankments). Nevertheless, after the 1990s, the policy for creating green spaces ended due to financial reasons, after reaching the figure of 7 m^2/inhabitant.

---

19 Available at: https://www.lesfabriques.fr/decouvrir/.

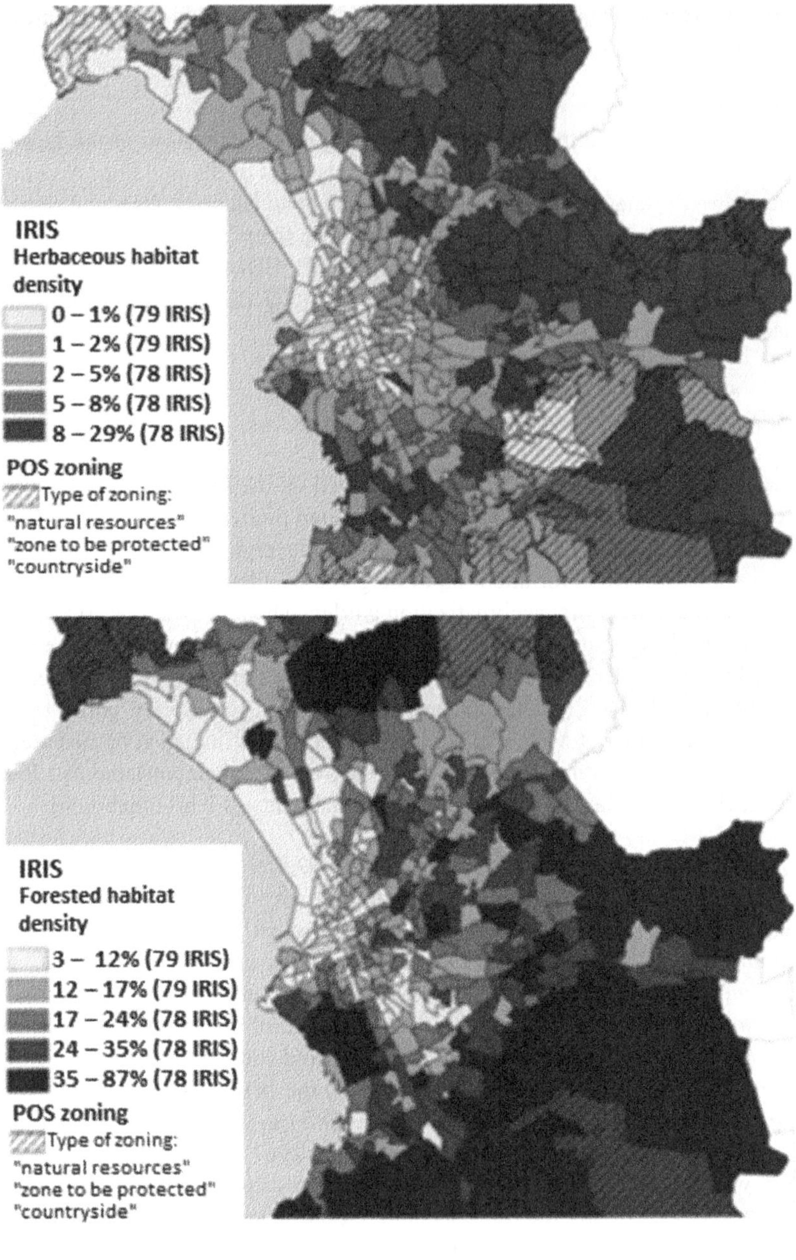

Figure 7.18. *Source: (Saint Martin 1998)*

In 1995, the goal of 10 m^2 per inhabitant was relaunched. A map of the distribution of public green spaces then revealed real disparities as to the distribution of green spaces across different neighborhoods. However, in 1998, 1031 green areas were identified, including around 400 gardens spread all over the city.

Today, if we refer to the figures published[20], over the 24,000 hectares which constitute the municipal territory, 14,000 hectares are urbanized and 10,000 hectares offer protected natural areas. One can identify 1,000 hectares of urban green spaces and 9,000 hectares of natural spaces located on the outskirts of the city. *Intra muros*, 605 hectares of urban green spaces are divided into public parks (for 800,000 inhabitants), country house gardens (private), and (private) collective gardens, green avenues. Therefore, the green spaces of the city of Marseille ultimately cover more than 700 hectares: parks, gardens, green street spaces, cemeteries and squares, and keep multiplying regularly.

Marseille has 58 parks with an area greater than 1 hectare, including 16 of more than 5 hectares. This heritage is supplemented by a network of approximately 273 squares and gardens. These spaces are also equipped with 132 playgrounds, something which necessarily raises the question of their management (Rouadjia 2017).

7.4.4. *But great precariousness*

Despite the creation of recent urban parks (*Parc du 26ème Centenaire, Bougainville*), the question of the amount of green space per inhabitant[21] has still not been solved. It is now also a question of integrating the analysis of the ecological quality of these spaces and their potential continuities.

In 2012, using the study of the Urban Green Network as a starting point, several Marseille researchers (LPED and LIEU laboratories) carried out a diagnosis of the state of Marseille's "green" areas. This took place in a contradictory context, torn between the local desire to promote urban development and the declared preservation of natural spaces, in the context of the enhancement of the Calanques National Park (created in 2012). The researchers then posited the question as to how the implementation of the Green Network should be interpreted, included in the SCoT PADD (*Projet d'Aménagement et de Développement Durable*) of 2012.

Their strategy consisted in bringing out the potential of nature present in the urban tissue of Marseille by relying on the notion of natural space (ENC, *Espace à Caractère*

---

20 Available at: http://www.marseille.fr/vdm/environnement/nature-en-ville/la-nature-marseille-presentation.

21 As compared to the reference of 10 m^2 of green space/inhabitant.

Naturel (Clergeau 2007)). Using the BD Carto and BD Topo databases, as well as data from municipal services (chloroville database), they produced a "reversed" cartography of the city, revealing the natural characteristics of the urban territory.

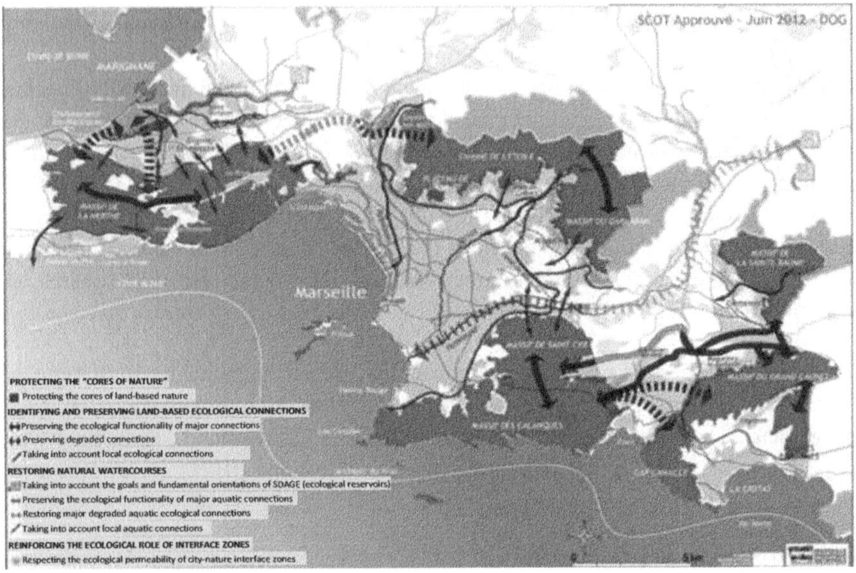

Figure 7.19. *Building an ecological framework. Source: https://www.marseille-provence.fr/index.php/documents/publications-scot/documentation/728-scot-mpm-cahier-dog/file*

Three units emerged from it. A coastal strip, comprising the port and the dense areas of the center, consisted mainly of rows of trees, public gardens and private gardens in the heart of the islet. The intermediate territory, still called "*Terradou*", because it was home to the former local agricultural suburb, presented a loose tissue which was largely transformed during the *Trente Glorieuses* (The Glorious Thirty, the 30 years following the end of the Second World War, a period of growth in France). Made up of village districts, wasteland and abandoned sites and public and private gardens of larger sizes, this territory still hosted ancient and varied vegetation. Finally, the peripheral, protected hills, mainly composed of scrubland, were the real city's biodiversity reservoirs.

This ecological structure of the city was also reinforced by the blue corridors of Huveaune, Jarret and Ruisseau des Aygalades.

While the SCoT PADD identified the large peripheral natural spaces and superimposed a blue grid on the existing water networks, it nonetheless remained very vague as to the qualification of the urban tissue itself. In this context, a Green Plan was also produced by AGAM. However, these documents paid little attention to informal vegetation, ecological continuities and the coherent management of these spaces on a city scale. In order to consider the city as a complete ecosystem, it was necessary to first understand its ecological functions and to go beyond the mere collection of cumulative green spaces.

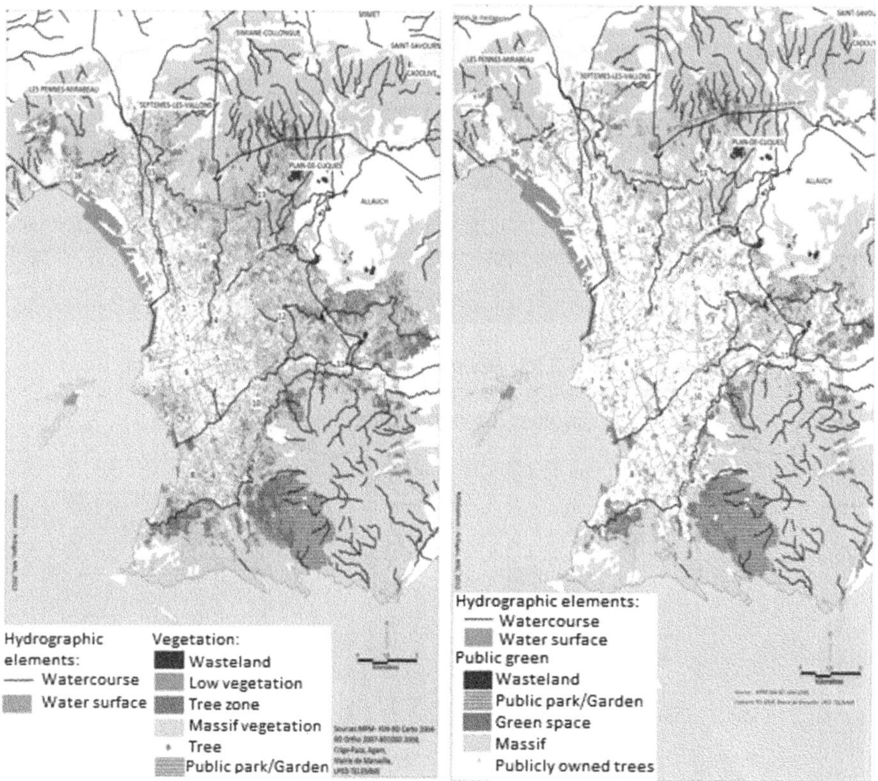

Figure 7.20. *Source: (Consalès et al. 2012)*

7.4.5. *What role should be given to the Marseille canal?*

Identified as belonging to the blue grid in the 2012 MPM SCoT PADD, the Marseille Canal is still used today to supply the city with drinking water. It is currently managed by the *Société des Eaux de Marseille Métropole* (SEMM), which ensures its proper functioning, operation and distribution. As it is likely to represent a danger for neighboring populations (10 drowning events in 30 years[22]), at present it is largely inaccessible and fenced in all its northern and eastern sections. It was even the subject of landfill work in the 1980s in order to optimize its management (Savine, Trois Lucs).

Figures 7.21 and 7.22. *Buried canal, La Savine, 15th district; Buried canal, St Antoine, La Martine, 15th district (photos: Séverine Steenhuyse)*

Figures 7.23 and 7.24. *Canal, Vallon de la Mûre, 14th district; Canal, Promenade du Cavaou, Plan de Cuques (photos: Séverine Steenhuyse)*

22 Available at: https://www.laprovence.com/article/edition-marseille/3456253/le-canal-de-marseille-10-morts-en-trente-ans.html.

Located in the outskirts, its layout is sometimes inserted into a dense and inhabited tissue, but in several sections it also presents dense and high-stemmed vegetation which helps to produce a real cooling urban micro-climate, and a "biodiversity buffer" space near the city's natural reserves.

In this context, the Marseille Canal does seem to be of particular interest. Even if today it seems difficult to imagine its transformation into a green leisure space, it naturally constitutes an axis for cooling and could act as an ecological transition corridor as regards the peripheral hills. Therefore, its presence plays an essential role in the conservation of ecological amenities in an urban context in which the outskirts continue to be densified. At this stage, its development should *at least* be based on a diagnosis of the ecological state of its banks and their existing extensions, or should be created in the nearby urban tissue. With this in mind, an inventory is being carried out on the whole of the Marseille route[23], which takes into account the identification of access, the urban contextualization, the quality of the environment and the existing ecological richness.

Figures 7.25 and 7.26. *Canal and view, Vallon de la Grave, 13th district; Covered canal, Domaine du Roy d'Espagne, 8th district (photos: Séverine Steenhuyse)*

7.4.6. *Conclusion: a natural space subject to multiple urban constraints*

Oscillating between a poorly supervised and very liberal urbanization, and the lack of concrete studies on the reality of the existence of an urban green network and its sustainability possibilities, it is difficult to imagine the local implementation of a UGN valuation policy, even if certain elements were notified in the PLUi (decree of June 28, 2018, coming into effect at the beginning of 2020).

23 As part of a long-term research action that I am leading in this field within the project[s] at ENSA-Marseille laboratory.

In addition, the local "green schizophrenia" (Consalès 2012) identified by researchers seems unlikely to contain and moderate local urban pressure. However, in the case of Marseille, the loose fabric of the outskirts embodies both the support for real ecological quality and the promise of preserved cool spaces. What could become of this basin city if its ecological breathing space gradually disappeared? Moreover, what is the perspective offered to these outskirts in the context of urbanization development that is self-contained[24] (Dorier and Dario 2016) and therefore little concerned with ecological and human continuity?

In this context, the thermal and ecological regulation services offered by the Marseille Canal seem to constitute an interesting long-term hypothesis for preserving quality "green and blue" spaces on the immediate periphery of the dense city center. However, its acknowledgment within the framework of the urban project directly raises the question of the temporal and spatial scales it must be integrated into during the planning phase. Considered as a structure embedded into a green grid which, in turn, would contain it within a blue grid, it could constitute the structuring axis for a set of easements concerning contiguous plots. Its influence and the challenges of its management would then become the opportunity to serve as a model of reflection on long-term urban planning through the invention of new ways of envisioning urban planning.

Figures 7.27 and 7.28. *Canal, St-Marcel, 11th district; Canal, Les Roubauds, Cuques/Allauch Map (photos: Séverine Steenhuyse)*

24 Available at: http://www.pensonslematin.fr/les-residences-fermees-a-marseille/ and especially Elisabeth Dorier, Julien Dario. Closed residences in France, chosen and built margins: Case study: Marseille, a residential closure laboratory (Grésillon *et al.* 2016).

7.5. Conclusion: design and build the city of tomorrow by integrating the issue of urban heat regulation as well as the revegetation of still available spaces

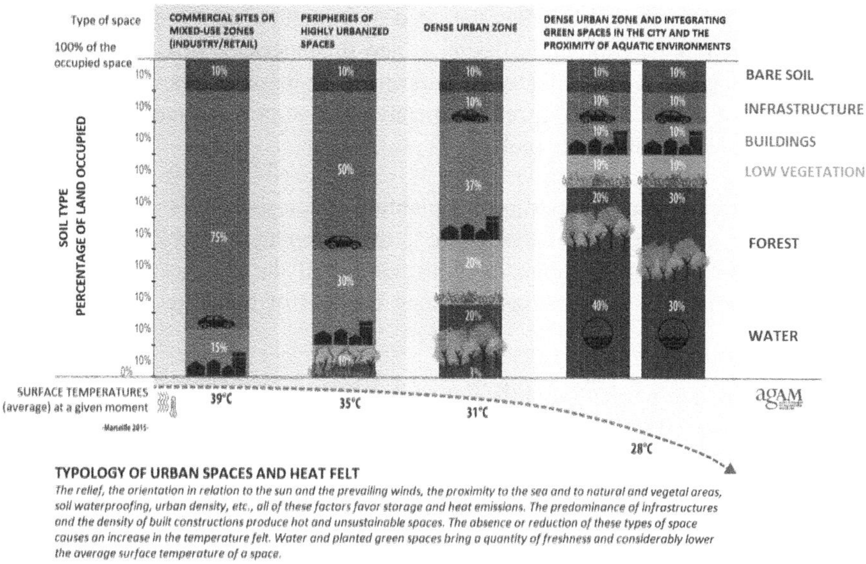

Figure 7.29. *Typology of urban spaces and heat felt. Source: (Regards ENVIRONNEMENT 2018)*

The combination of two public policies from different sectors makes it possible to reveal both their potential complementarities and the difficulties for their respective implementation. The management of urban heat islands presupposes the design of a new urban tissue reconsidering the proportion and distribution of occupied spaces and those left free according to a new model, while integrating a planting policy.

As for the Urban Green Network, its implementation supposes a strong political mobilization capable of carrying a real project of controlled densification and structuring of urban tissues along "green axes", which are hardly perceptible and poorly recognized today, especially in the loose spaces of the city.

However, in view of the observation of the limitations identified within the framework for the implementation of these two public policies, the question is still open as to how to concretely formulate the model and to achieve this sought-after, virtuous and sustainable city.

7.5.1. *Rethinking the definition of the value of spaces*

The reintroduction of natural spaces, and more generally, plants in the city, seems to remain one of the most effective solutions to fight against the urban heat island effect[25]. However, it implies the need to introduce empty spaces without a specified economic function in the design of new urban spaces, which induces a new way of thinking about the collective enhancement of these spaces. In addition, the preservation of empty spaces in recent operations must be combined with a policy encouraging planting in private plots.

Public greening can be based on the planting of transport axes (malls, avenues, railway lines, alignment plantations, etc.), the greening of public lands (green spaces, outdoor spaces of public facilities, etc.), and the enhancement of private green land (perimeter of buildings, greening the feet of buildings, green facades, etc.).

Two sectors are particularly relevant to the city of Marseille: commercial areas and social housing (HLM type) parks.

7.5.2. *The local planning document and its concrete implementation*

At this stage of the process, the PLU/PLUI seems the most adequate document to formulate this new attention required by "urban voids". Subject to environmental assessment under Article R 121-14 of the Urban Planning Code, the latter must also explain its consideration of the environment and the measures envisaged to avoid, reduce or compensate for the negative impacts of planned development. In this perspective, it must also define monitoring indicators making it possible to analyze the environmental consequences of its impact on space consumption.

As a regulatory measure, it offers a set of possible protection zones: A (Agricultural) and N (Natural) zones, reserved locations, and landscape preservation goals (Art L 123-1-5 7°), real-estate control goals (Art L 123-1-5 8°), ecological preservation (L123-1-5.9°), graphic transfer of ecological continuities (Decree of February 29, 2012), classification of certain elements under EBC (Art L 130-1), definition of cycle paths and greenways and identification of Ecological Interest Areas (EIA) to be protected.

25 We refer to the work of Marjorie Musy (2012).

7.5.3. *The Marseille case*

The Marseille situation is based on a double heritage characterized by a large surface area of a poorly-valued former industrial and highly-mineralized wasteland near the city center and a tradition of ultra-liberal and poorly-supervised urban development.

In addition, the city has incomparable assets related to its multiple interactions with interior natural spaces (the heart of islets) and peripheral spaces (landscapes, loose tissue, quality of vegetation). In this context, the infrastructure of the Marseille Canal constitutes a unique opportunity to create a real axis of peripheral transverse urban transition which combines the functions of a potential green-blue grid to associate a set of ecosystem benefits with it that are intended for neighboring populations.

The question comes down to considering the possible terms for this implementation.

Finally, if we take up the question posited at the start of this chapter, even if the expectations related to the reduction of heat islands and those related to the realization of UGNs initially seemed to join, it nonetheless seems difficult to make them interact for their respective implementation. On the one hand, they do not effectively concern the same urban spaces: the dense city center for the UHI, and the periphery for the UGN. On the other hand, they do not exactly share the same development strategies: operational urbanism for the UHI and planning for the UGN.

The question of real-estate control remains, in all cases, the key to possible solutions on different types of tissues to be treated: the design of new neighborhoods, which must acknowledge the challenges of ecological transition through the production of new urban models, the intervention in existing neighborhoods, which must integrate a policy for planting available spaces, improved soil drainage and introducing cooling mechanisms, and finally, loose peripheral tissues, which must be partially protected in virtue of their biodiversity. These last tissues should become the first objects of attention of the authorities and public policies in particular, in order to preserve their qualities in the long term, while sparing transitions towards large bordering natural spaces.

7.6. References

ADEME (2012). Guide de recommandation pour lutter contre l'effet d'îlot de chaleur urbain : à destination des collectivités territoriales. ADEME.

ADEUS (2014). Les îlots de fraîcheur dans la ville. Les notes de l'ADEUS.

APPA Nord-Pas de Calais (2014). Végétation urbaine. Les enjeux pour l'environnement et la santé. Thematic dossier, APPA Nord-pas-de-Calais.

Arrif, T., Blanc, N., Clergeau, P. (2011). Trame verte urbaine, un rapport Nature – urbain entre géographie et écologie. *Cybergeo*: *European Journal of Geography*, 574 [Online]. Available at: https://doi.org/10.4000/cybergeo.24862.

Bertaïna, J., Riou, J., Belmont, L., Lemaire, A., Carre, G. (2012). La Trame verte et bleue dans les Plans locaux d'urbanisme. Methodological guide, DREAL Midi-Pyrénnées.

Blanc, N., Glatron, S., Schmitt, G. (2012). Trames vertes urbaines : recherches en sciences humaines et sociales. *Développement durable et territoires*, 3(2) [Online]. Available at: http://developpementdurable.revues.org/9370.

Borruey, R. (1983). *Marseille bastidaire : le rôle des bastides dans la structure morphologique du territoire péri-urbain de Marseille*. École d'architecture-Marseille, Marseille.

Bowler, D.E., Buyung-Ali, L., Knight, T.M., Pullin, A.S. (2010). Urban greening to cool towns and cities: A systematic review of the empirical evidence. *Landscape and Urban Planning*, 97, 147–155.

Briche, E., Collange, L., Hubert, M. (2017). *Climat et ville : interactions et enjeux en Provence-Alpes-Côte d'Azur*. GREC-PACA, Marseille.

Brisepierre, L., Castiau, M., D'hermy, A. (2013). Marseille, nature en ville et espaces publics. Atelier Pédagogique Régional, ENSP/EPA Euroméditerranée.

CETE Méditerranée (2012). Mise en œuvre de la trame verte et bleue dans les Bouches-du-Rhône. Report, CETE Méditerranée, Aix-en-Provence.

Collectif (2017). Trame verte et bleue : les outiles pour sa mise en oeuvre. Cahier technique no. 91, Montpellier AFB, Montpellier.

Consalès, J.N., Goiffon, M., Barthélémy, C. (2012). Entre aménagement du paysage et ménagement de la nature à Marseille : la trame verte à l'épreuve du local. *Développement durable & territoires*, 3(2) [Online]. Available at: http://journals.openedition.org/developpementdurable/9268.

Dacheux-Auziere, B. (2018). Quelles modalités d'écologisation de la pratique des paysagistes-concepteurs ? L'exemple des projets de parcs publics de ces 4 dernières décennies. Le cas de 5 parcs marseillais. PhD Thesis, Larep, ENSP.

DEEV (1995). Guide des parcs et jardins de Marseille. Guide, DEEV, Marseille.

Dorier, E. and Dario, J. (2016). Closed residences in France, chosen and built margins: Case study: Marseille, a residential closure laboratory. *Pensons le matin* [Online]. Available at: http://www.pensonslematin.fr/les-residences-fermees-a-marseille/.

Feix, I., Marquet, S., Thibier, E. (2017). *Aménager avec la nature en ville*. ADEME, Angers.

Grésillon, E., Alexandre, B., Sajaloli, B. (2016). *La France des marges*. Armand Colin, Paris.

Hagel, Z. (2013). Ville durable : des concepts aux réalisations, les coulisses d'une fabrique urbaine. Marseille ou l'exemple d'une ville mediterraneenne. PhD Thesis, LIEU, Aix-en-Provence.

Héloïse, M., Lapray, K., Musy, M. (2017). Diagnostic de la surchauffe urbaine. Méthodes et applications territoriales. Research study, ADEME.

Kervadec, T. (2012). Mettre en œuvre la Trame verte et bleue en milieu urbain. Note, Etd, Centre de ressources du développement territorial.

Le Bot, J.M. and Philip, F. (2012). Les trames vertes et bleues urbaines, un nouveau support pour une cité verte. *Développement durable et territoires*, 3(2) [Online]. Available at: https://doi.org/10.4000/developpementdurable.9318.

Linglart, M., Morin, S., Paris, M., Clergeau, P. (2016). Méthodologie de mise en place d'une Trame verte urbaine : le cas d'une communauté d'agglomération, Plaine Commune. *Cybergeo: European Journal of Geography* [Online]. Available at: http://journals.openedition.org/cybergeo/27713.

Lyon Confluence (2019). Lyon Confluence, laboratoire urbain de la ville de demain. Press release [Online]. Available at: https://www.lyon-confluence.fr/en/publications/lyon-confluence-urban-testing-ground-city-tomorrow-2019.

Marseille Informations (1976). Priorité aux espaces verts. Report no. 78, Marseille Informations, Marseille.

Millennium Ecosystem Assessment (2015). Guide to the Millennium Assessment Reports [Online]. Available at: https://www.millenniumassessment.org/en/.

Musy, M. (2012). L'étude des microclimats urbains : champ de recherche à l'interface entre climatologie, urbanisme et génie-civil. *VertigO – la revue électronique en sciences de l'environnement* [Online]. Available at: http://journals.openedition.org/vertigo/11841.

Regards ENVIRONNEMENT (2018). Chaud dehors ! De la fraîcheur face aux îlots de chaleur urbains. Publication no. 75, Agam, Marseille.

Rodriguez-Potes, L. (2013). Milieux urbains et ambiances thermiques : influence des arbres sur le microclimat urbain selon les paramètres de plantation et leur impact sur le confort thermique. Cas des villes d'Aix-en-Provence et de Marseille. PhD Thesis, ENSAM, Marseille.

Roncayolo, M. (2002). *Lectures de villes. Formes et temps*. Parenthèses, Marseille.

Rouadjia, A. (2017). Le paradoxe de la gestion des espaces verts : entre volonté de maîtrise et laissez-faire. *VertigO – la revue électronique en sciences de l'environnement* [Online]. Available at: http://journals.openedition.org/vertigo/18338.

Saint Martin, Y. (1998). Les espaces verts publics à Marseille : bilan d'une politique pour un meilleur environnement urbain. *Méditerranée*, 89, 87–90 [Online]. Available at: https://www.persee.fr/doc/medit_0025-8296_1998_num_89_2_3056.

Soares, I., Vandroux, B., Magalon, N. (2010). Lutte contre les îlots de chaleur urbains. Référentiel conception et gestion des espaces publics. Reference guide, Grand Lyon Communauté Urbaine, Lyon.

Terrin, J.-J. (2015). *Villes et changement climatique : îlots de chaleur urbains*. Parenthèses, Marseille.

Toublanc, M. and Bonin, S. (2012). Planifier les trames vertes dans les aires urbaines : une alliance à trouver entre paysagisme et écologie. *Développement durable et territoires*, 3(2) [Online]. Available at: http://developpementdurable.revues.org/9347.

Valette, E. and Cordeau, E. (2010). Les îlots de chaleur urbains. Répertoire de fiche connaissance. Study, IAU IdF, Île-de-France.

Vidal-Naquet, P. (1993). *Les ruisseaux, le canal et la mer*. L'Harmattan, Paris.

Ville de Marseille (1976). Guide vert de Marseille 1975–76. Guide, Marseille.

8

Summer Comfort in the Face of Climate Change: Design Adaptation and Rehabilitation

This chapter refers to the lessons taught at the Architecture School of Toulouse in relation to the studies conducted at LRA[1] and whose subject is "Summer comfort in the face of climate change: Design adaptation and rehabilitation".

Thousands of dwellings have been built and renovated with low-energy, in compliance with various thermal regulations and many housing facilities which have been rehabilitated thanks to government subsidies. These new constructions and rehabilitations have responded to the energy issues, focusing on reducing energetic precariousness and decreasing energy expenses during the winter. But the results over the last four years are irrefutable. The purchase of air conditioners, which could reach 5.6 billion in 2050 worldwide, is increasing considerably in France, revealing a mishandling of summer comfort.

Many questions are raised, such as the real reductions in energy expenditures, the solutions considered for combining summer and winter comfort, as well as new housing rehabilitation which is not efficient every season. The research carried out at the LRA on this issue originates in the reflection exercised during the lessons, namely: what design can adapt to climate change, and what rehabilitation solutions can be adopted? The contents and teaching proposals such as technical and socio-economic knowledge modules, have evolved and an environment seminar has

Chapter written by Isabelle VERVISCH-FORTUNÉ.

For a color version of all the figures in this chapter, see www.iste.co.uk/aithaddou/transition.zip.

1 LRA: Architecture Research Laboratory at Toulouse.

been added to the curriculum. The first part of the chapter will explain this reflection and methodology. The second part will report on the research carried out, the way in which the lessons articulate around this issue, focusing on heritage, and overall cost of project as related to ecological transition.

8.1. Introduction

As regards the architect's role in ecological transition, this chapter deals with their training and its evolution in recent years. How should future designers be trained at architecture schools to address the issue of climate change? What are the commitments made within ENSA Toulouse to instantly integrate these changes into training?

Confronted with significant issues and the reality of these changes in recent years, the courses provided in Toulouse integrate the subject of ecological transition throughout the curriculum, at different scales. These teachings, backed by research carried out at the LRA, are evolving to better take into account the changes taking place.

In the Midi-Pyrénées region and more particularly in Toulouse, for the past 15 years housing facilities have been built in accordance with regulations. Likewise, housing facilities have been renovated into low consumption modes, with BBC (*Bâtiments Basse Consommation*, Low energy buildings), BEPOS (*Bâtiments Energie Positive*, Energy-plus buildings) or *Effinergie* labels in Midi-Pyrénées so as to match the various thermal regulations. While these new achievements and renovations have perfectly responded to the energy issue aimed at eliminating energy poverty as well as at reducing energy expenditure during winter and use renewable energies, evidence for the past three to four years is indisputable. The latest studies carried out reveal a significant increase in temperatures and relative humidity, causing discomfort for occupants in insulated and airtight buildings. It is therefore urgent that we change the way in which our homes, neighborhoods and cities are designed. This is why modules are offered not only during the initial training, but also during continuing training in order to help students and architects design differently, taking into account the various regulations and the evolving climate change.

This article opens up with the current context which leads us to adapt the teachings' contents every year, so as to closely match evolving needs. Then, we present the different research work areas conducted on this theme as well as the accompanying teachings, which feed on this progression. Finally, three Master's courses related to ecological transition taught at ENSA in Toulouse are presented. We will discuss the lessons to be developed and the links with research that should be further examined. Lastly, we will conclude with the perspectives to be developed in research as well as for the training of architecture students to encompass ecological transition.

8.2. The current context

In order to cope with the increase in energy expenditure, mainly winter and heating-related, and to eradicate energy poverty, thermal regulations have been enforced since 1974. They have continued to evolve with the RT 2000, which added a requirement for summer comfort, and with the RT 2005, taking into account bioclimatism and renewable energies (Collet 2011). RT 2012 has become more demanding of expected results and the means of implementation. In recent years, 2,000 new homes have been built per year in Toulouse complying with the various BBC, *Effinergie*, etc. labels. At the same time, the government has encouraged energy renovations through various subsidies. Despite this, the number of renovations is decreasing because thermal regulations are too expensive and deter a large number of users who only find long-term savings. The latest low-consumption buildings compliant with the RT2012 thermal regulations in force provide an answer to winter comfort[2]. And the results are convincing. Of course, this is the first step to address the problem, namely eliminating energy "sieves", improving winter comfort and encouraging eco-citizen actions.

But, climate change is not a utopia. It is responsible for the increase in temperatures, especially in the southern regions of France. For two consecutive winters, temperatures have not been negative in Toulouse. In addition, the last few summers have had higher temperatures for longer than five consecutive days. This is already affecting the environment and populations.

At the same time, discomfort has been noted during the summer period; especially in recent BBC buildings, such as those that were both constructed after the Second World War and that are isolated. In order to protect oneself against the ever increasing heat due to dense urbanization, the recommended solutions – bombarded with publicity – have focused on cooling the premises with the focus on air conditioners. The purchase of air conditioners could reach 5.6 billion in 2050 worldwide and is incessantly increasing in France (AIE 2018).

In addition to defacing architectural heritage, these solutions go against the main goals of the Ministry of Ecology and Solidarity Transition: to reduce household energy consumption (Michel 2017). Currently, RT2012 requires buildings to have reduced energy expenditure, and in response to this, many buildings are constructed with insulation and without air circulation[3]. To date, the only rule in terms of

2 Texte officiel RT 2012 ; Décret du 26 octobre 2010 relatif aux caractéristiques thermiques et à la performance énergétiques des constructions (JO du 27 octobre 2010).
3 Texte officiel RT 2012 ; Décret du 26 octobre 2010 relatif aux caractéristiques thermiques et à la performance énergétiques des constructions (JO du 27 octobre 2010).

summer comfort is the Tic (*Température intérieure conventionnelle*, Conventional Interior Temperature).

We therefore call into question the response to this new climate problem and the air conditioning equipment that can be found on the façades of buildings in Toulouse and many other cities in southern Europe, such as Tirana, in Albania (Figure 8.1). Do we wish to be in the same situation in the next few years?

Figure 8.1. *Façade of a building in Tirana (photo credit: Isabelle Vervisch-Fortuné)*

What can we say about the energy savings achieved? What comfort is there for the occupants? It was through these observations that we became interested in summer comfort for buildings, and we used this questions as a starting point. Now we ask: how do we design or rehabilitate buildings so that they meet the need for summer comfort?

In this regard, several questions emerge, among which we can mention: *are (ecological/energetic) transitions only a question of savings?*

Since their inception, thermal regulations have been no exception to the rule: all articles are based on primary energy savings. So what about summer energy costs related to air conditioning?

So, *is comfort to be considered merely a question of insulation?* Thermal regulations bring forward the role of insulation as a solution to every problem.

The current design of many buildings and the solutions envisaged do not make it possible to regulate the interior temperature without resorting to air conditioning. What are the reasons for this discomfort? How should we deal with it? While the goal of reducing energy costs in winter has been achieved, is this compatible with summer comfort? What responses can be offered?

Are we going to continue to design and build these buildings (Figure 8.2)? They are the answer currently provided by the regulations, even though we are already aware of their limitations.

Figure 8.2. *New dwellings in the Purpan district (31) (photo credit: Isabelle Vervisch-Fortuné)*

In the face of climate change and user comfort, the architect is eco-responsible and co-responsible, together with other construction players. To face this challenge, architects need to be trained as thoroughly as possible and must continue to look for new designing solutions respecting the human being.

The research carried out at the LRA and the teachings go in this direction:

How do we adapt design and energy rehabilitation to this change in climate? What lessons are provided in order for us to combine both comfort, economy and respect for the environment and the user?

8.3. Method

The issue of housing comfort is one of the current research axes at Toulouse Architecture Research Laboratory, which helps to support the training of architecture students. Not only does it encourage contributions and reflection, but it also prompts the evolution of teaching contents. It is complementary to a more global reflection. Each certification grants that the teachings have been reworked to match professional reality as closely as possible, and that they anticipate future skills for architects. Research brings content and reflection to the teachings. Likewise, the work carried out in the project workshops, in the teaching of technical knowledge, raises questions that make research take a leap ahead.

Since I arrived at the Toulouse School of Architecture in 2008, two syllabuses have been offered, largely influencing part of the content, and focusing intently on ecological transition. Even at that time. The Toulouse school was already known for being inspired by the values of architecture and the environment. Courses on sustainable development were available, as well as project workshops on the environment.

Two main areas are leading the development of education and research: the preservation of heritage and successive thermal and environmental regulations.

The research carried out from 2010 to 2013 jointly with the Ministry of Ecology, Energy, Sustainable Development and the Sea and the Scientific and Technical Center for Building mainly focused on winter comfort. In order to respond to this issue, teachings were implemented during the 2010–2016 certifications. Table 8.1 presents an overview.

Semester	Teaching/Content
Semester 1	Sustainable architecture: environment, sustainable materials
Semester 3	Thermal: fundamentals
Semester 6	Energetic: project application using ArchiWIZARD software
Semesters 7/8/9	Project workshop and seminar

Table 8.1. *Teachings available during 2010–2016 addressing the topic of ecological transition*

These courses and supervised practical works made it possible to enrich the curriculum and the project with architectural responses related to the environmental problem, on the one hand, and concerning thermal regulations, on the other. A corpus of buildings was studied in order to feed the research work and accompanying theoretical contributions. These first courses enabled the student to approach design by integrating regulatory constraints from the very beginning. We have thus designed BBC, BBIO, HQE, etc. buildings. For this, the learning of digital tools and software has been integrated into the new syllabus in direct relation with the teaching of technical knowledge. In semester 6, for example, a TM (Teaching Module) is offered comprising three teachings: Project construction principles, Project energy, and ArchiWIZARD learning. These three teachings have been intertwined and aligned for consistency (see ENSA 2016 syllabus).

At the same time, projects on vernacular materials (Hygroba 2013) are carried out at the Architecture Research Laboratory. They focus on rehabilitation projects, and the question of ensuring balance between technical performance modifications, building sustainability and the reasonable evolution of the architectural form. A first course proposal on energy diagnosis and rehabilitation was offered in 2014, enriched by research work.

For three years, the BBC projects constructed have revealed their limitations in the Toulouse region, especially during the last summers when the indoor temperature was around 30°C with a high humidity of over 80% for several days. BBC design as it used to be taught at the Toulouse School of Architecture may no longer be suitable for all seasons. New research on the topic of summer comfort is being carried out on old buildings, as well as more up to date studies on BBC buildings. It is on the basis of ongoing research in relation to environmental and comfort issues, that training at the school of architecture has evolved and now offers new courses based on the Research, Environment and Architecture axis in the 2016–2021 authorized syllabus.

One of the main goals of the new syllabus is the pooling and transversality of teachings. Likewise, projects have become multidisciplinary, cross-breeding the challenges so that the student is confronted with the reality of architectural and technical design.

The teachings on ecological transition presented in Table 8.2 are backed up by the research work carried out at the LRA (Architecture Research Laboratory) where I participate, but the list is not exhaustive.

S1 → M1	Toca Tierra workshop participation (Figure 8.3)
S1	Construction and sustainability
S3–S6	Construction and energy
S7	Is raw earth the answer to summer comfort? What's the current concept?
S7–S8	Seminar transition(s) in question: environment, landscape, architecture
S7–S9	Heritage workshop project
S9	Energy rehabilitation of vernacular buildings
S9	Construction economics (overall cost of energy rehabilitation)

Table 8.2. *Teaching topics offered in 2016–2021 covering ecological transition*

Figure 8.3. *Students working on adobe manufacture*

8.4. The research/teaching duality for energy transition

This part of the chapter is devoted to the work carried out at the Architecture Research Laboratory and the teachings directly related to the research axis I am engaged in. Other research projects on this theme are being developed at the laboratory at different scales (the urban scale, in particular), making it possible to pool skills and enrich research. They are not developed in this contribution.

8.4.1. *Research work on energy transition*

For approximately the last 10 years, my work at the Architecture Research Laboratory has mainly focused on the energy rehabilitation of vernacular buildings. They are set within the "Architecture, Landscape, the Environment" field. The studies are carried out at several scales, ranging from matter to materials, from building to blocks. The first studies through the Hygroba project focused on the hygrothermal energy of vernacular materials and different ways of insulating. They were employed as communication support media and notebooks for the Ministry of the Environment (Hygroba 2013). This made it possible to highlight the qualities of old buildings, and to understand the hygrothermal balance of old walls associated with contemporary insulating materials (Vervisch-Fortuné and Valkhoff 2013). The teaching of building old wall insulation, by favoring the perspiration of the wall, has been integrated into the courses. However, only the results have been used to enrich the content of the courses. For the time being, students cannot use the Wufi software, which slows down the progress for exploiting results, which therefore remain qualitative. This work was carried out jointly with the CETE of Strasbourg and *Maisons Paysannes de France*, providing a body of vernacular buildings. This prompted the creation of the Energy Rehabilitation course which feeds scientific reflection by means of its continuous development.

Jointly, studies on technical equipment were carried out using the ArchiWIZARD software which was integrated into the courses, in Semester 6, then in Semester 3, thus enriching the Bachelor's degree as well as the Master's degree courses.

Several other reflections have been carried out on energy rehabilitation, often performed by self-builders. The question arose as to the architect's role. After several appraisals of buildings rehabilitated in self-construction mode, it appeared that the training of self-builders was one option, but that the support of an expert such as an architect really was a prerequisite for any rehabilitation. The self-builder could benefit from agencies such as Anah, which offer aid and ensure quality rehabilitation and heritage safeguarding (Vervisch-Fortuné 2014).

In particular, thermal regulations emphasize the energy performance and energy savings that energy rehabilitation can generate. Concretely, research focused on the real savings that users could obtain from insulating their housing. It appears that home insulation is more of a civic response to improve comfort rather than one that saves money. For a large number of citizens, sometimes undergoing energetic precariousness, investment takes 30 to 50 years to recover, depending on the kind of work undertaken. State assistance makes it possible to reduce this period but it requires an important investment all the same (Vervisch-Fortuné 2015). This economical question is at the heart of energy transition and represents an important part of my research studies, having started my professional life as a construction economics engineer. It is a driving force for its deployment. Like any high-performance, renewable technical equipment, solar equipment has been studied in order to understand the lack of enthusiasm and the decline in its attractiveness. A study was presented at an international conference in Tirana, Albania, on the best compromise to estimate between investment and energy savings (Fortuné 2018). To approach these works of an economic nature, a corpus of buildings is studied following a specific study plan (diagnosis, planning proposal, thermal and energy study with the help of software and savings study). This plan is included in the teaching offered in the Master's degree. Work has been tackled on current urban densification and its consequences at a building scale.

On the one hand, the latest research focuses on raw earth (a material found in many vernacular buildings in the Midi-Pyrénées) and summer comfort. On the other hand, it explores the overall cost of construction and rehabilitation in the face of climate change. The first study in progress is being developed concurrently in the classroom as well as at the laboratory, highlighting the study of materials such as raw earth due to its vernacular characteristics and its comfort performance. Raw earth appears as an interesting hint to combine summer comfort with winter comfort. This leads us to reflect on the possible rehabilitation of new BBC housing – already obsolete – in the face of climate change.

The second study deals with the additional cost of construction or rehabilitation caused by the rise in temperatures. Two axes are privileged. The first one concerns the cost of the implementation and the costs of maintaining and operating air conditioning. The second one concerns the plus/minus value related to "bioclimatic" design and the choice of materials.

The research performed at the LRA on the theme of energy transition feeds the reflection carried out in the Master's courses, particularly on the question as to which design better adapts to climate change, and which rehabilitation solutions can be adopted. They evolve according to the current events, and help enrich the proposals for new teaching modules, as well as contents, like those presented below.

8.5. The evolution of Master's degree courses

Even if throughout the bachelor's degree, students have courses which are directly related to ecological transition, such as sustainable materials, thermal comfort, etc., it is especially at the Master's level that students can fully integrate these concepts into their projects.

The three examples of teaching modules presented below attest to the work carried out in the second cycle on the issue of energy transition. These modules complement one another and are offered to students as different choices. As explained in the previous section, they are backed up by research axis in architecture, landscape and environment, finding inspiration in it and enhancing the work carried out.

8.5.1. *Rehabilitation of vernacular buildings*

This elective course is taught in Semester 9. It proposes to study a vernacular building *in situ*, both from the energy and heritage point of view. The students are confronted with the reality of the work carried out in the case of energy rehabilitation. Every year, students study new buildings at a particular site. For three years, students carry out both a site analysis and a survey in the project workshop. This enables them to perform the diagnosis, the proposal development and the energy study during the 40-hour module. In addition, by attaching this module to the project workshop, the student discovers more cohesion between the project and the teachings that nourish it.

They carry out architectural (Figure 8.4), structural and constructive, environmental, energy and sanitary diagnoses. Above all, these are qualitative diagnoses to understand the building's state of preservation, as well as its construction principle. The energy diagnosis is carried out using the ArchiWIZARD software, which makes it possible to obtain an order of magnitude of energy needs.

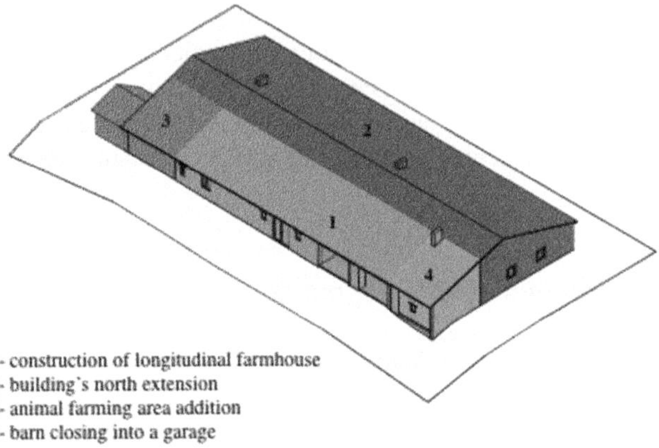

Figure 8.4. *Axonometry: building evolution, historical stages (source: S95 student's example)*

From all the information gathered, they make a layout proposal of the building for two different uses: daily use as accommodation and occasional use as community premises. The architectural and constructive diagnoses enable them to make a coherent proposal in terms of constructive typology, the ordering of the facade's elements, and the respect for heritage. Structural and sanitary diagnoses (Figure 8.5) reveal the disorders which need to be corrected before one proceeds with any energy improvement intervention.

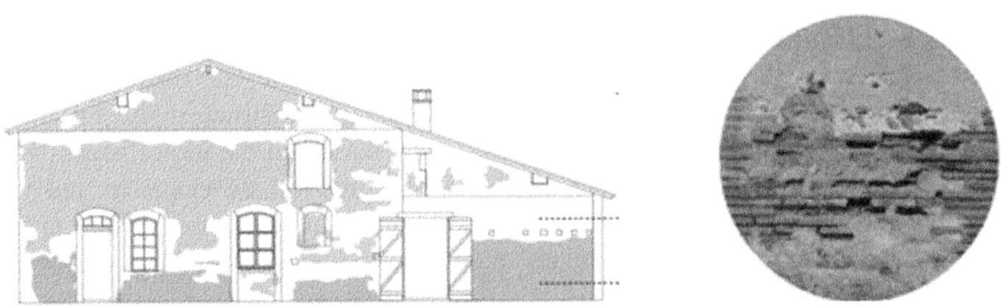

Eastern facade

Close-up of the condition of the facade

Element	Description	Sanitary condition/ Consequent pathology	Cause for pathologies	Proposed solution
Plasters, Eastern facade	Lime plaster: white on the walls and pink around the window frames 80% conservation on the upper part of the wall 10% conservation on the lower part of the wall	Highly degraded	Capillary rise of water and its imprisonment between the brick and the cement plaster	Stripping of degraded plasters and replacement with lime plaster

Figure 8.5. *Sanitary diagnosis excerpt (source: S95 student's work)*

The student's layout proposal is supplemented by an energy study.

CALCULATION OF THE POWER REQUIRED FOR THE PELLET BOILER

	Housing	Public equipment	Total	Required performance in kW		Quantity of granules in t	Storage volume in m³
Heated area with heating	162 m²	269.8 m²	431.8 m²	Calculation according to the site elyoterme.fr	41.45	For 1 kw, we need between 250 to 350 kg of granules	ρ = 690 kg/m³
Heated volume per radiator	142.2 m³	232.3 m³	374.5 m³	Temperate climate with good insulation (35 W/m³) = 13107.5 w	13.11		
				Total	55 kW	13.75 t	20 m³

SIZING OF THE CAST IRON RADIATORS

Room volume (m³)			Required performance (moderated climate with good insulation)	Type of radiator			
				Number of elements	Height (mm)	Length (mm)	Emission (w)
Housing	Bathroom	27.8	973	7	750	546	1008
	Bedroom 1	62	2170	8	750	624	1152
				8	750	624	1152
	Bedroom 2	52.4	1835	13	750	1014	1872

Public equipment	Workshop 1	100.5	3517.5	13	750	1014	1872
				13	750	1014	1872
	Workshop 2	69.8	2443	9	750	702	1296
				9	750	702	1296
	Workshop 3	62	2170	15	750	1170	2160

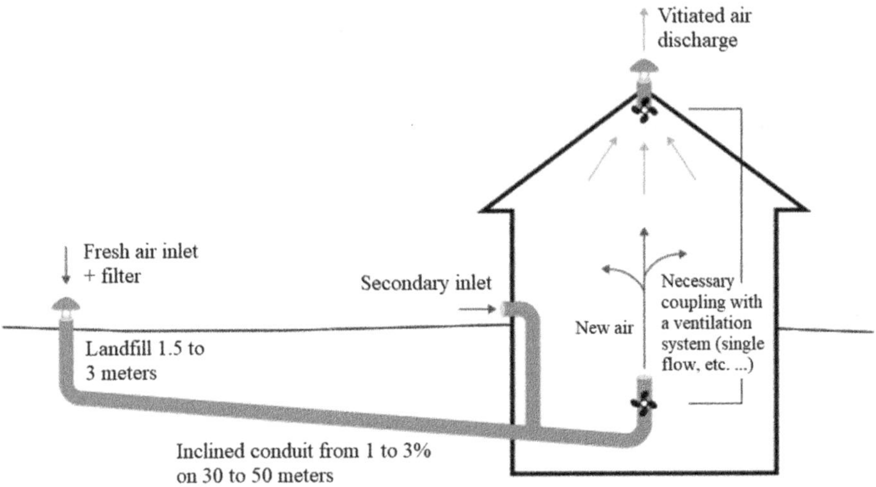

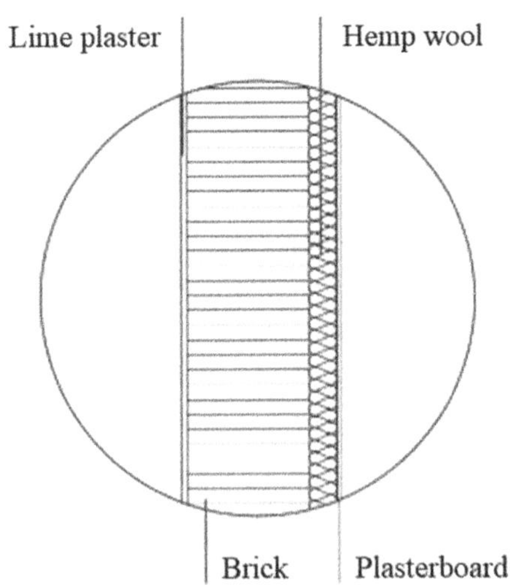

WALLS

Thermal resistance to be reached: R = 2.3 m^2K/W			
	Depth e (m)	**Thermal conductivity λ (W/mK)**	**Thermal resistance (m^2K/W)**
Loadbearing member: *brick*	0.42	0,60	0,70
Insulation: *hemp wool*	0.10	0,038	2,63
Exterior coating: *lime plaster*	0.02	0,70	0,03
Interior coating: *plasterboard*	0.013	0,25	0,05
Film resistance of wall (Rse + Rsi)			0,17
		R eff	3,58

Figure 8.6. *Proposal of ecological technical equipment (source: S95 student's work)*

In terms of energy efficiency, three levers need to be considered:

– The first one, called passive energy efficiency: concerns the reduction of needs by improving wall thermal resistance in relation to the environmental problem, by promoting solar gain in winter and protecting against it in summer.

– The second one, called active energy efficiency: concerns the improvement of technical equipment and the use of renewable energies (Figure 8.6).

– And the third one, user behavior: a reflection is carried out by the students in order to clearly define the uses attributed to each space and to take into account their proposal, the attitude as well as the habitat's self-appropriation performed by each user. The definition of a behavioral model associated with various consumptions and levels of comfort/discomfort made it possible to identify a long-term relationship between behavior and comfort levels (Batier 2016). This teaching is supplemented by the construction economics module, which enables the student to define energy needs and their impact in economic terms.

8.5.2. *Raw earth as an answer to summer comfort*

This course is set in Semester 7 in the form of a 15-hour intensive course. Its goal is to make the student aware of and reflect upon the current issue of building comfort, so as to face increasingly hot summers. The role of raw earth is studied to supply better summer comfort. It comes down to understanding the constructive typology of this vernacular architecture. It is studied through projects in need of rehabilitation, in order to understand their physical, technical and environmental characteristics.

Method

Several buildings are studied. The purpose of this is to reconcile summer and winter comfort while respecting heritage and reducing energy expenditure. One of the houses studied and discussed is located in Cuq-Toulza in the Midi-Pyrénées, in a rural environment (Figure 8.7).

Figure 8.7. *Raw earth dwelling in Cuq-Toulza in Midi-Pyrénées (photo credit: Isabelle Vervisch-Fortuné)*

The few hours devoted to this module only make it possible to tackle a qualitative study of the building. This is only a first step. Further complementary studies will make it possible to quantify energy expenditure and to determine summer indoor temperatures. First, the owners were questioned about their way of life, their feelings about their housing, for which an inventory was carried out afterwards. The idea is to reconnect with the well-being in ancient buildings during hot summers, understanding underlying reasons and suggesting rehabilitation proposals.

Results

In this case, the interview with the owner was very informative. Feeling genuine devotion for this 18th century farmhouse, its owner was interested in heritage preservation as well as the potential of its vernacular materials. This resulted in the removal of some plasterboard partitions in the ground floor, replaced by raw earth brick walls. The new cob partitioning (Figure 8.8) upstairs made it possible to redistribute the space.

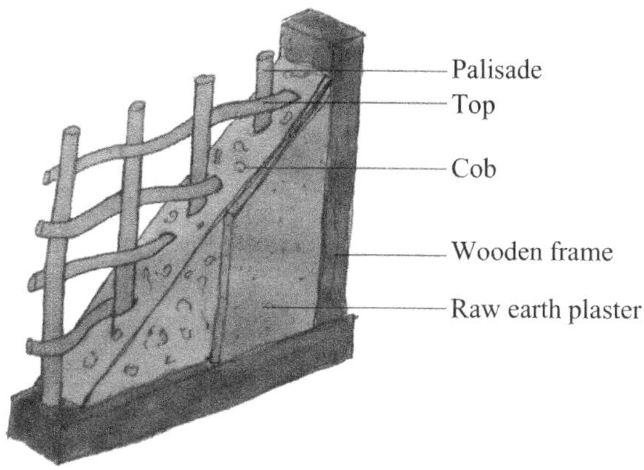

Figure 8.8. *Sketch of the cob wall (source: S7 students)*

A constructive, structural and sanitary diagnosis suggested traces of humidity and a loss of adhesion of the raw earth cross-wall.

The problem with this building is humidity and it is absolutely necessary to let it breathe. The configuration of the house, with a large central corridor, and the possible circulation between each room, favors good ventilation, which should definitely be kept. Volumes, generous but not too large, are very pleasant. During the visit, it was very hot outside, and the house, open on all sides, was cool. The owner instructed them about thermal comfort, but also mentioned the acoustic comfort of raw earth.

The peripheral walls are made of stone and raw earth, with a thickness of 70 cm, as for the interior walls, these are made of raw earth brick, their thickness being 50 cm (Figure 8.9). Their inertia promotes the regulation of interior temperatures.

Based on the set-out principles, in addition to the inertia of walls and their insulation, the other response to energy efficiency is a proposal for technical equipment made in relation to winter and summer comfort (Figure 8.10). Here the question is to carry out a reflection contrary to what is currently done in new constructions, where everything is insulated to protect it against the cold. Taking summer comfort into account leads us to reconsider circulation, the building's natural ventilation, protection against sunlight, insulation and the building's design itself. For example, the conservation of buffer spaces present in old buildings, such as under-roof attics, which may or may not be habitable.

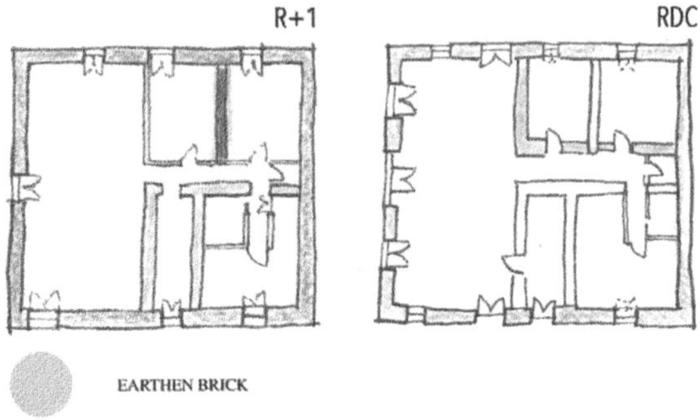

Figure 8.9. *Interior raw earth and cob walls (source: student work S75)*

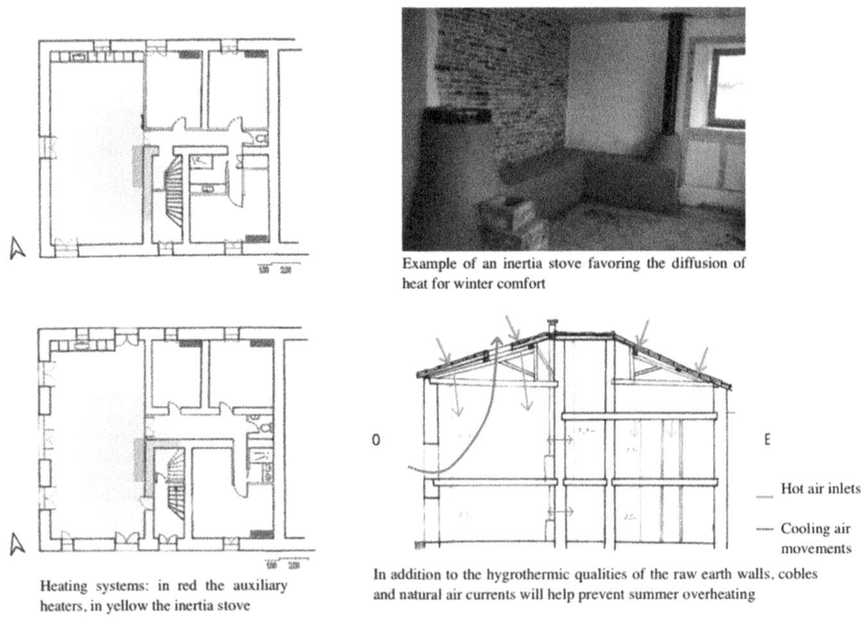

Figure 8.10. *Proposal of technical and ventilation equipment (source: S75 student's work)*

Discussion

Like many other buildings, this earth and stone construction is analyzed in view of understanding how to build contemporary vernacular architecture today, or to rehabilitate heritage constructions. Faced with climate change, in front of the abundance of so-called "thermos" housing to meet various regulations, this building analysis provides the student with knowledge about the expectations in terms of comfort, feelings and a representation of the self within a habitat.

After a diagnosis and an observation of the building as well as an interview with the inhabitants, students are able to propose an improvement of the habitat in terms of thermal comfort for all seasons. Through on-site analysis, they discover the important role of materiality in comfort, as well as effusiveness.

They learn about the main principles for rehabilitating or designing a comfortable building in the Toulouse region, where temperatures can be really high in the summer, with very high relative humidity. Other comparative studies, using ArchiWIZARD, make them aware that the rules applied to limit heat loss are perfectly applicable to winter comfort, but opposed to summer comfort. They can learn a lot from the design of old buildings.

8.5.3. *Design of a new raw earth building*

This S7 optional student course took place in 2017. Research carried out in the laboratory focuses both on new low-consumption buildings as on vernacular buildings. In this module, we are interested in the new design of a building for warmer climates. We ponder about the current design of buildings in response to Toulouse's climate, for example, and find inspiration in the design of vernacular buildings.

Method

Based on the reading of different earthen buildings, from the Gauzin-Müller exhibition (Gauzin-Müller 2017) and a variety of theoretical contributions, the students make a design concept proposal for a contemporary building in raw earth, meeting the new requirements of energy expenditure reduction and home comfort for all seasons. In 15 hours of supervised practical works, they only have time to sketch a project by explaining its main guidelines.

Results

On the one hand, the main guidelines of a project include the building's volumetry, inspired by farm typologies in Midi-Pyrénées (Figure 8.11). The parallelepiped is underlined by a single-slope roof to differentiate itself from a more modern construction, composed of two dwellings.

Figure 8.11. *South facing facade (source: S75 student's work)*

An advance is deployed on the northwest facade and denotes the entrance to each accommodation.

The land extends to the south-east towards spaces made up of deciduous trees, making it possible to shade the dwellings in summer and both to heat up and let light pass through in winter (Figure 8.12).

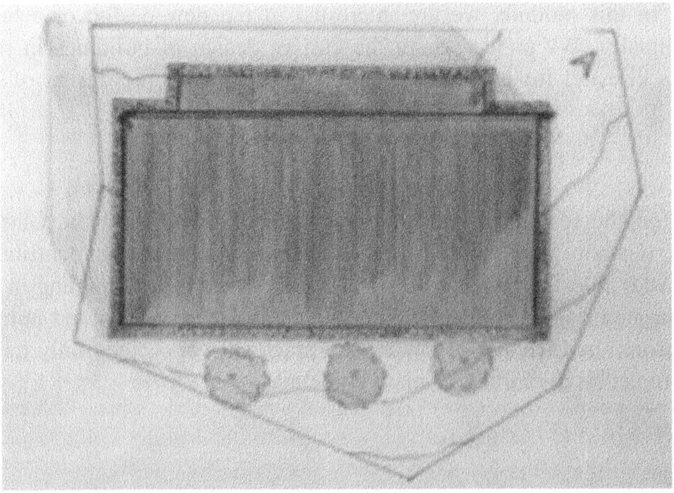

Figure 8.12. *Ground floor (source: S75 student's work)*

On the other hand, the south facade makes it possible to benefit from optimal lighting, therefore saving energy and using free resources, for heating purposes in winter and protection in summer (Figure 8.13).

Figure 8.13. *Schematic section (source: S75 student's work)*

To the north, the installation of a glass ribbon makes it possible to benefit from constant light. The double height enables air circulation, communicating with the windows to the north as well as constant natural regeneration (Figure 8.14).

Figure 8.14. *Natural ventilation (source: S75 student's work)*

The architectural and structural approach of this duplex adobe building highlights the choice of the durability of structural elements, as well as the use of natural elements (sun, air, vegetation).

The building rests on a basement at chest level in terracotta, a local material, to fight against the capillary rise of water coming from the ground (Houben *et al.* 2006). The complex which was put in place includes a stone bedding for draining purposes around the building. The walls to the west, exposed to bad weather (heavy rain in the southwest facade) are made of solid brick, and protected by roof overhangs. The east and south facing facades are made from adobe and protected by eaves. The choice of an adobe composition for these facades is guided by several factors: its regulatory capacity, its capacity to capture heat or freshness, and to restore these with a 10-hour delay (Figure 8.15).

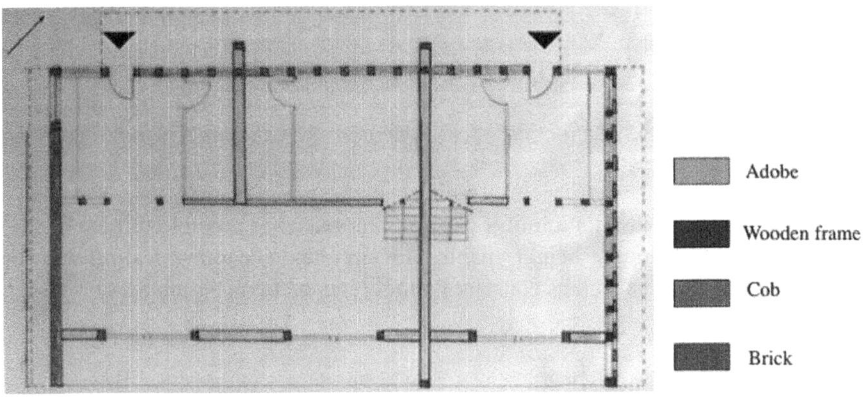

Figure 8.15. *Wall materiality plan (source: S75 student's work)*

The first floor is covered in window panes in order to optimize sunlight and ventilation. In addition to wood for floors and partitions, cork is also offered for the sake of recycling and is accessible at a lower cost. The roof extends beyond the walls to create protection from the sun in summer and to let the sun in during the winter, as well as to prevent water from flowing onto the facade, which causes the resulting degradation of materials. The frame is supported by wooden studs integrated into the walls, thus avoiding pressure on the adobe walls and to avoid any perforation. They take the load off the frame and transfer it directly to the ground.

A whole reflection is carried out on the work of the east facing facade in solid brick, making it possible to create a perforated wall on the house's exterior, marrying graphic and visual design, letting light pass through, while protecting the house from the eastern wind.

Discussion

The aim of this exercise is to discover a different design from the conventional one, respecting the environment, user comfort, materials, taking into account the characteristics of each material and their implementation. Buildings that are pleasant to live in while also protecting the inhabitants from the hotter climates can be designed for the future. This teaching feeds on the work carried out on Mediterranean architecture (Gauzin-Müller and Corset 2016). This work, still at a sketching phase, deserves to be pursued in depth, endowing it with more substantial modules and longer study hours.

8.6. Energy theories and practices to be reinforced

Research work and Master's degree teachings in the field of energy echo one another, pooling skills and making it possible to prompt a leap ahead in scientific data, while enriching the field of Architecture Sciences and Techniques. For several years, the Toulouse National School of Architecture has set up courses dedicated to the environment in correspondence with the research developed at Grecau laboratory. Both research and teachings have diversified to better adhere to and anticipate the climate change that is taking place. Since 2009, teachings have been offered on how to take thermal regulations into account and reduce energy expenditure when considering the design of any architectural project. They are based on the various research projects carried out within the Architecture Research Laboratory.

As research evolves, the courses offered differ from one certification to another, making it possible to correlate scientific data with professional reality.

The hours allocated to research in architecture schools, like the hours given in the field of Architecture Sciences and Techniques, do not favor great extension for making proposals. Deeper involvement in project workshops could favor this development. Although this has already been initiated in semesters 7 and 9, timeline considerations have blocked this momentum. The next syllabus, which is already being redesigned, must continue the path initiated by providing technical and socio-economic knowledge, and further integrating this knowledge within projects.

In order to guarantee consistency in ecological transition contributions and to integrate them into projects, the current curriculum is structured as follows.

The Bachelor's degree cycle (semesters 1 to 6) is devoted to fundamental theoretical contributions from experiments valuing the acts of "understanding through experience" and "learning from mistakes". They are supported by the use of digital tools to encourage students in their approach and thinking processes. These teachings, calling on several themes and skills, are pooled to help the student in understanding phenomena or conceptual choices. For example, in Semester 1: Construction and Sustainability, the choice in materiality induces a particular typology and a specific type of architecture, which must simultaneously acknowledge environmental considerations, the intrinsic characteristics of the material and its implementation.

Within the project developed in Semester 6, the student is encouraged to reflect on the energy proposal by choosing the type of insulation that meets "sustainability" considerations, energy savings and its implementation. The student confirms their intuitions using the ArchiWIZARD software. A workshop on raw earth, with the presence of Amaco, is offered to all students over a week in order to rediscover this local, healthy material with high thermal inertia.

The Master's degree is designed as an in-depth study, offering the student several study paths to choose from. The different teaching modules are developed through a common project. These modules offered in semesters 7, 8 and 9 complement one another and resonate to help students understand the complexity of a project that requires the successful combination of architectural design, technical design, heritage preservation, energy and material savings, and the environment. For example, a student interested in ecology can choose the following curriculum (Table 8.3).

Semester 7	Design TM: Project workshop: Heritage in progress Theories and Culture TM: Matter and materials Seminar TM: Transitions in question (Architecture, Landscape, Environment) Technical knowledge TM: Raw earth and comfort
Semester 8	Theories and Culture TM: A material history of construction Seminar TM: Transitions in question (Architecture, Landscape, Environment)
Semester 9	Design TM: Project workshop: Heritage in progress Technical and socio-economic knowledge TM: Construction economics Introductory module TM: Energy rehabilitation or: Research Curriculum

Table 8.3. *Example of a curriculum chosen at the Master's level*

It is through these courses that the question of winter and summer comfort, which today seems opposable, is addressed. Bachelor's degree courses on energy will no longer be able to deal with only one type of insulation or thermal bridges, but rather will have to integrate the fact that it is not advisable to always insulate everything. It is necessary to calculate the total annual expenditure and the overall cost taking into account building cooling. In this case, the so-called "thermos" building is no longer economical in the Toulouse region and its conception, just like its materiality, needs to be redesigned.

8.7. Conclusion

Faced with the observation of discomfort in new BBC buildings, research has focused on this problem over the last four years. Ongoing studies are already showing the limitations of an insulated and airtight building during the summer season. The 2012 thermal regulations contributed to the design of energy efficient buildings for the winter period. But deprived of inertia, in the summer season, these airtight buildings consume the energy gained during the winter. The new courses take this development into account and provide answers on how to design healthy and economical buildings for all seasons. We can no longer only take into account winter comfort, by increasingly insulating and airproofing buildings. New teaching modules are offered to expose the new research to students and to lead them to new thinking and conception of their future project, in the face of ongoing climate change. This of course involves the rehabilitation of old buildings and the use of vernacular materials. But also new courses, such as a Material History of Construction, are offered to relearn the know-how of buildings from the past, which is key if we hope to respond to summer comfort. The solutions envisaged must take into account the specific characteristics of the material, such as effusiveness. We must rethink the design including buffer spaces, and put aside the attic's layout. Talking about ecological transition without acknowledging the overall cost and the feeling of being in one's habitat is not in line with the current challenges. While it is true that the Master's degree courses have evolved, these notions will have to be initiated as early as they are in the Bachelor's degree. It will also have to include digital diagnostic and energy performance tools within the project. Finally, although the rehabilitation of 20th century buildings is on the syllabus, it is necessary to plan to rehabilitate the buildings of the 2000s, with the aim of improving summer comfort and making them more convincing from an economic point of view.

8.8. References

AIE (2018). Le coût écologique de la climatisation. Report, International Energy Agency, Paris.

Batier, C. (2016). Confort thermique et énergie dans l'habitat social en milieu méditerranéen : d'un modèle comportemental de l'occupant vers des stratégies architecturales [Online]. Available at: https://hal.archives-ouvertes.fr/tel-01324038/file/These_Batier_2016.pdf.

Collet, P. (2011). La réglementation thermique de 1974 à aujourd'hui. La RT2012 bientôt en ordre de marche. *Actu-Environnement* [Online]. Available at: https://www.actu-environnement.com/ae/dossiers/energie-batiment-rt2012/historique-reglementation-thermique.php4.

Fortuné, I. (2018). What is the better compromise between energy savings and investment in energy rehabilitation? Conference, S-Arch 2018, Venice.

Fortuné, I. (2019). Is the energy performance aimed today in buildings compatible with summer comfort? What type of construction is most suitable? International Conference in Greece, Athens.

Gauzin-Müller, D. (2017). *Architecture en terre d'aujourd'hui*. MUSEO, Toulouse.

Gauzin-Müller, D. and Corset, F. (2016). *Architecture méditerranéenne d'aujourd'hui. L'intelligence collective pour mieux bâtir*. Editions Eyrolles, Paris.

Houben, H., Guillaud, H., CRATerre (2006). *Traité de construction en terre*. Editions Parenthèses, Marseille.

Hygroba (2013). CETE de l'est, ENSA/LRA de Toulouse. LMDC et maisons paysannes de France. Hygroba, étude hygrothermique de la réhabilitation des parois anciennes [Online]. Available at: http://lra.toulouse.archi.fr/lra/activites/projets/hygroba.

Michel, L. (2017). Plan Climat ; Ministère de la transition écologique et solidaire ; Paquet solidarité climatique : quatre mesures concrètes pour les foyers modestes. Report, Gouvernement français, Paris.

Vervisch-Fortuné, I. (2014). Remaking housing policy in France: How to accompany the self-builders in the energy rehabilitation of vernacular buildings. ENHR International Conference, Edinburgh.

Vervisch-Fortuné, I. (2015). Housing policy in France, is it focused on people's needs? What real economy the user does by insulating its housing? ENHR International Conference, Lisbon.

Vervisch-Fortuné, I. and Valkhoff, H. (2013). A solution to the housing crisis: Rehabilitation of old buildings both from a hygrothermal and conservation point of view. Study on the hygrothermal behavior of existing walls associated with insulation. ENHR International Conference, Tarragona.

9

Mutations Project at ENSA Nancy

In 2015, at the École nationale supérieure d'architecture in Nancy (ENSA Nancy), a team of eight teachers, researchers and practitioners from different disciplinary fields was created in response to the many changes in the contemporary world (environmental, social, economic, etc.). The team, named "Mutations", intends to bring together teachings (particularly architectural design), scientific mediation actions (the "*Rencontres Interdisciplinaires – Mutations*"), and eventually, common research projects. This chapter presents the team's framework and then makes an initial assessment[1] of its activities, based on confrontation with the issues raised by contemporary change, all within a collaborative and multidisciplinary perspective.

The chapter covers its subject in three stages. First, we discuss the history of the founding of the Mutations team and its constitution, as well as its current research axes; then, the organization of highlights called *Rencontres Interdisciplinaires Mutations* (Mutations Interdisciplinary Meetings) (two of which have already taken place); and finally, a range of teaching activities including the Master's degree project workshops.

Chapter written by Cécile FRIES-PAIOLA.

For a color version of all the figures in this chapter, see www.iste.co.uk/aithaddou/transition.zip.

1 Curien and Fries-Paiola (2018) provide an exhaustive list of the activities carried out by the team during its first two years.

9.1. Origins of the team – the Mutations project

9.1.1. *Concerns shared in 2015*

In 2015, several teacher-researcher-practitioners from ENSA Nancy established that they shared concerns about three phenomena:

1) the gravity and acceleration of the environmental crisis (climate change and the collapse of biodiversity as the most worrying (Grandcolas 2017), as well as social and economic concerns (the rise in inequalities (Wieviorka 2017)) and the profound cultural transformations accompanying them (in particular, the impacts of digital technology);

2) the difficulty in finding the right perspective and tone to discuss these complex and anxiety-provoking questions with students (Pelissolo 2019), and in finding the best way to try to collectively provide answers to them (facing the challenges of reducing environmental impact and adapting to the new ecological context);

3) the lack of integration of these questions, albeit essential among architecture students, even though the impact of these ongoing and future changes will particularly affect their practice and being aware of their role as levers for action[2], in view of the professional position they will assume.

These shared concerns made it possible to bring together a small nucleus of teachers-researchers-practitioners, determined to reflect on a common action project, simultaneously addressing three important issues (the first one, thematic, the other two methodological):

1) questioning the evolution of architectural practices and their meaning, in relation to the above-mentioned contemporary ecological issues;

2) the treatment of these questions from a multidisciplinary perspective, with particular emphasis on Human Sciences[3];

3) the need to structure the fields of teaching, research, practice and scientific mediation, so that they can enrich one another.

---

2 "The construction sector represents 44% of the energy consumed in France, by far outnumbering the transportation sector (31.3%). Every year, the construction sector emits more than 123 million tons of Carbon Dioxide" in 2017, according to the Ministry for Ecological Transition. Available at: https://www.ecologique-solidaire.gouv.fr/energie-dans-batiments (last accessed on March 7, 2020).

3 For further information on ecologically-oriented human sciences, see for example Bird Rose (2019); and for the outlines of these environmental humanities and the approaches they propose, see for example Choné *et al.* (2016).

9.2. Gradual team building

This founding nucleus decided to organize a first study day in 2016, open to all teachers from the school sharing these concerns and wishing to reflect upon a common project. This first opportunity for exchange enabled everyone to present their work, thoughts and perspectives, as well as giving rise to the opening of a first Mutations project workshop that very same year.

Far from being fixed, the team gradually expanded thanks to a second study day in 2017, leading to a fruitful collaboration in parallel between the Laboratory of the History of Contemporary Architecture (LHAC) – to which the researchers of the team are attached – and exchanges with the school's teaching community, particularly newly recruited teachers.

Today, the team is made up of eight members, with various profiles but all concerned with the treatment of the relationship between environmental issues and contemporary practice of architecture:

– Christelle Chalumeaux (architect, teacher-researcher ATR, LHAC);

– Émeline Curien (DEA holder, Doctor of Art History, teacher-researcher TPCAU, LHAC);

– Cécile Fries-Paiola (DEA holder, Doctor of Sociology, teacher-researcher SHSA-VT, LHAC and 2L2S-UL[4]);

– Hervé Gaff (DPLG holder, Doctor of Philosophy, teacher-researcher SHSA, LHAC);

– Serge M'Boukou (HDR, Doctor of Anthropology, teacher-researcher, LHAC);

– Mathias Rollot (architect, Doctor of Architecture, teacher-researcher TPCAU, LHAC);

– Claude Valentin (architect, teacher-researcher ATR);

– Christian Vincent (architect, teacher-researcher TPCAU).

9.3. The link between teaching, research and practice

The Mutations project therefore proposes to combine teaching, research, practice and the mediation of architecture. The development of the Mutations project also led the team to structure action that can be broken down into three points:

---

4 Le Laboratoire Lorrain de Sciences Sociales (2L2S), University of Lorraine, France.

1) the development of research guidelines related to questions of current and future environmental and social changes;

2) the organization of events (*Rencontres Interdisciplinaires Mutations*), combining education, research, architectural practice and mediation;

3) the organization of project workshops integrating the thematic (environmental changes, etc.) and methodological (multidisciplinary, etc.) aspects of the Mutations project.

The first point will be briefly outlined here, before tackling the other two points in each of the following sections. Thus, in the context of a global reflection on the structuring of the LHAC, the consideration and study of these mutations has become the general framework of one of the three axes named, "Production and reception of architecture" – also a highly multidisciplinary one, integrating the works of researchers in philosophy, anthropology and sociology.

Today, the research of various members focuses on the evolution of practices and the representations related to these changes. Let us mention a few examples. Claude Valentin works on the link between architectural practice and research, by integrating CIFRE (*Convention industrielle de formation par la recherche*, Industrial Training through Research Agreement) within his agency[5] or conducting research-actions, in particular on the topic of short supply chains (such as wood in the Vosges region). Émeline Curien works on the analysis of contemporary architectural productions implementing broader ecology (such as that of Gion A. Caminada (Curien 2018)). By means of a sociological approach, Cécile Fries-Paiola calls into question the representations of the environment–human relationship carried by practitioners of alternative habitat conceptions (Fries-Paiola 2017).

The research team is still being set up these days and the challenge is to be able to better integrate – both within ENSA Nancy but also with teacher-researchers from other ENSA institutions – in order to implement joint research and respond to calls for projects related to the themes within our scope of action.

9.4. *Rencontres Interdisciplinaires Mutations* (Mutations Interdisciplinary Meetings)

The first reflections within the framework of the Mutations team led us to gradually reconsider the place and organization of the team within the LHAC but also to organize an event (which has already taken place on two occasions, one in December 2017 and the other in November 2019): the *Rencontres*

5 HAHA Architecture workshop, see: www.haha.fr.

Interdisciplinaires Mutations (RIM). In fact, within the framework of teaching and research, on which we have been working so far, there was a feeling of lack of space for the expression and production of ecological thought that could invite both students and the teaching community to be not only bolder in relation to the complex state of our contemporary world, but also more alert and able to imagine alternatives in the fields of architecture/territories.

Therefore, the spirit of these meetings is to promote an open and conscious vision of architectural research and practice, in the face of ongoing major ecological, socio-economic, political and cultural changes and those yet to come, as well as to provide a multidisciplinary inventory of contemporary changes for a wider audience, including not only students, researchers and teachers, but also architectural practitioners. This knowledge about major contemporary issues, which are not always directly understood by the many players in architecture, is now a necessary prerequisite for informed action. It must therefore become one of the major challenges for teaching and carrying out research in architecture.

9.4.1. *The goals of the event*

The RIM meetings intend to meet three main goals: scientific, educational and cultural. The first goal is to invite researchers from disciplinary fields outside architecture, including *a priori* distant fields, which deal with an aspect of contemporary change, in order that the school's students and teaching team can benefit from precise insight into the questions addressed by these researchers (as was the case when the field of economics was included at the first event, or biology at the second). In return, we also wish to implement the favorable conditions for enriching guest researchers both by their mutual encounter, and by their confrontation with the world of architecture/territories.

The second goal is to expose students to specific pedagogical situations by confronting them with speeches and perspectives they are not usually exposed to. Thus, during the RIM meetings, the guests engage in the framework of a project workshop session, during which the students present a state of the art of their work, and are invited to discuss specific aspects of their research project, depending on the guests' disciplinary fields and research topics. They also take part in seminars supervised by teachers from the team, seminars which are prepared and conducted by students.

Finally, the last goal of these RIM meetings is a cultural one. These meetings seem to be a way of contributing to the visibility and acknowledgment of the numerous complex and accelerated changes we are confronted with, and will continue to be. Although the situation is urgent, we know the extent to which a change in practices will necessarily involve a change in representations; these events intend to contribute to this awareness.

9.4.2. *The program and the guests*

RIM 1 and 2 were organized according to an identical framework, which includes four components. The first part is educational, including guest interventions in the project workshop and during the seminars. The second part is devoted to the scientific exchange between the guests and the members of the Mutations team. The third part concerns architectural practice, including a practitioner's conference. Finally, the last section proposes to open discussions with a wider audience, approaching scientific mediation in the form of a roundtable which brings together the various guests and concludes the meetings.

For the first event, Dominique Bourg (philosophy professor, University of Lausanne), Ali Douai (lecturer in economics, University of Nice Sophia Antipolis[6]), Jean-Marc Stébé (sociology professor, University of Lorraine), and the architects Ivano Iseppi and Stefan Kurath were the guests for the meetings entitled, "An inventory to tackle action, from the global environment to local architecture", to which Agnès Sinaï (journalist) also contributed. For the second event, we invited Philippe Grandcolas (CNRS research director in biology, biodiversity[7] specialist), Thomas Le Roux (CNRS researcher in history), Pierre Janin (architect), Mathias Rollot (architect, doctor of architecture) and Catherine Deschamps (socio-anthropologist) for two days devoted to the question "Human, non-human, for architectures and territories of the living".

9.4.3. *First appraisal and perspectives*

The first RIM in 2018 was an opportunity to publish the work of the Mutations team conducted within its first two years, from 2016 to 2018. The publication of Curien and Fries-Paiola (2018) was the means for making the team visible and communicating the educational, scientific and mediation activities it organized: The

6 And a member of "*Économistes atterrés*", an association created in 2011 made up of researchers, experts and citizens, which aims to "stimulate collective reflection and public expression of economists who do not resign themselves to the domination of neoliberal orthodoxy" (http://www.atterres.org/page/association).

7 And director of UMR 7205 which brings together the ISYEB (*Institut de Systématique, Évolution, Biodiversité*), the National Museum of Natural History, the CNRS, Sorbonne University, the *École Pratique des Hautes Études* and the University of Antilles.

actions of RIM naturally hold a special place in this. The second of the RIM meetings, as well as the rest of the team's work, was set for publication in 2020.

Three main lessons can be drawn from this first RIM. First of all, the urgency and gravity of the environmental situation have been widely reported and documented, as have the current difficulties of our socio-economic and political system in dealing with it. But at the same time, solutions have also been identified, of which architects can be real bearers: for example, through the reflection on the "common good" by Ali Douai (Curien and Fries-Paiola (2018, pp. 69–86)).

Although interdisciplinarity seems obvious today – particularly within schools of architecture, a field that is broadly open to various disciplines – its value was brought back to us during these RIM meetings, as so many visions and speeches differ on certain issues and perspectives[8]. The modeling of different scenarios and their dissemination within the scientific community constitute major challenges in order to be able to collectively reflect about these mutations.

Finally, the interest sparked by this initiative among students encourages us to continue in this direction and to organize other meetings. They feel largely concerned by these changes, and are keen to enhance their knowledge and their thought using sources beyond the scope of the architectural field, with the aim of building a true professional standing. And this is one of the major difficulties in which we must always endeavor to better support them: How can we integrate significant and complex issues and challenges within a limited and localized project?

Thus, continuing to organize such meetings seems obvious to us, in particular by developing three lines of thought. First of all, it seems important to continue to broaden the disciplinary spectrum of guests: the confrontation of architecture with domains which appear remote *a priori* enabled students (and teachers) to understand new challenges for practice[9], or to think differently about previously identified

8 During the first of the RIM meetings, Dominique Bourg and Jean-Marc Stébé each spoke of changes through the prism of their discipline. While the first author addressed the topic of the inevitable reduction in world population given the resource limitations we are gradually reaching, the second author focused on population increase, mainly urban populations, due to the fact that in the most populated countries and/or whose population is growing more rapidly, urban transition has not been fully completed.

9 For example, this is the case of the intervention by Thomas Le Roux, who by developing the concept of pollution, revealed the need to adjust the perspective on the contemporary environmental situation, whose novel aspect – which sometimes determines the difficulty in taking action – has been put into perspective: in fact, the games of complex actor seeming to

challenges[10]. After these first two open meetings, which enabled us to confirm the interest in and the need for such a working format, it will be a question of considering more targeted themes for the upcoming meetings: these will be decided according to current affairs topics and the evolution of changes. Finally, it seems important to us to continue to work on the modalities and format of these meetings, by affirming and further articulating the different stages (pedagogy, research, practice and mediation) they propose.

9.5. Mutations project workshops

Above all, the Mutations team is an educational team: its activities are mainly structured around the organization of Master's degree project workshops, including a final thesis (PFE, *Projet de Fin d'Etudes*). After discussing the principles on which these workshops are based – which go beyond those around which the team was initially united – we will appreciate how the students came to own the framework offered to them.

9.5.1. *Framework and project workshop principles*

In order to build relationships with the world and decidedly ecological living spaces, two levers seem essential: to support students in gaining a broader vision of situations, by confronting multiple perspectives, and building up and reaffirming a committed, informed position in relation to a site or area.

The workshops are based on three main operating principles. First of all, these are connected to the research carried out by the members of the Mutations group, insofar as the seminars and courses offered in parallel emanate from these reflections; in fact, they are reinforced within the workshop through a few specific thematic courses. Then, the construction of a broad and complex perspective on a site or area seems crucial, which is why all the members of the team accompany the students during a workshop. The different architectural, technical, artistic, sociological, anthropological and philosophical points of view continuously come

---

slow down effective environmental action have emerged and settled since the beginning of the industrial revolution.

10 We are thinking, for instance, on the intervention by Philippe Grandcolas about biodiversity: while this question is not foreign to architectural and urban practice, its definition, its evolution, its relation to climate change, etc., are aspects which were explored and deepened during the seminar, providing us with a more complex view on this question.

together during the project's development. Finally, this method makes it possible to gradually guide the students towards greater autonomy in practice. They are encouraged to build their own approach, without having an appointed teacher, making the most of the different points of view of accompanying instructors.

For the workshop of the first semester in Master 2 – also conducted for a time with students of the first semester in Master 1 – the framework offered to students is that of a common site for the entire group, for which they have to envision its transformation by 2050, considering the current main environmental, socio-economic and cultural changes. They are free to develop the position that seems the most relevant to them, while imagining ways to mitigate environmental impact as well as those related to the adaptation to new living conditions.

As the feasibility study appears to us to be an essential training issue for the architectural profession, the workshop specifically focuses on this phase of the project. For the students, it is a question of constructing their problem: they are at the origin of the commission, and must master the challenges. A field survey, often leading to a meeting between various actors, is added to the conventional analytical tools used. It is following this dynamic that the project program gradually develops. Finally, a reflection carried out on the basis of different scenarios makes it possible to gradually build an individual (or collective) perspective on the intervention site.

In fact, the four areas already explored – varying from year to year – were thought out and constructed as typical situations, sparking a more global reflection. Workshops took place in four sectors of the Nancy urban area[11]: a large complex (in Laxou), a peripheral town (Malzéville), a peri-urban town (Seichamps) and a fragment of a rather dense urban fabric (in Nancy). Let us see some of the positions taken by the students within the framework of these different workshops.

11 Under the sphere of the prefecture of Meurthe-et-Moselle, Nancy is a student city with little over 100,000 inhabitants *intra muros*, and a little more than 435,000 inhabitants within its urban area. Its metropolis, Grand Nancy, is made up of 20 municipalities, all of which are part of those explored within the framework of the workshops.

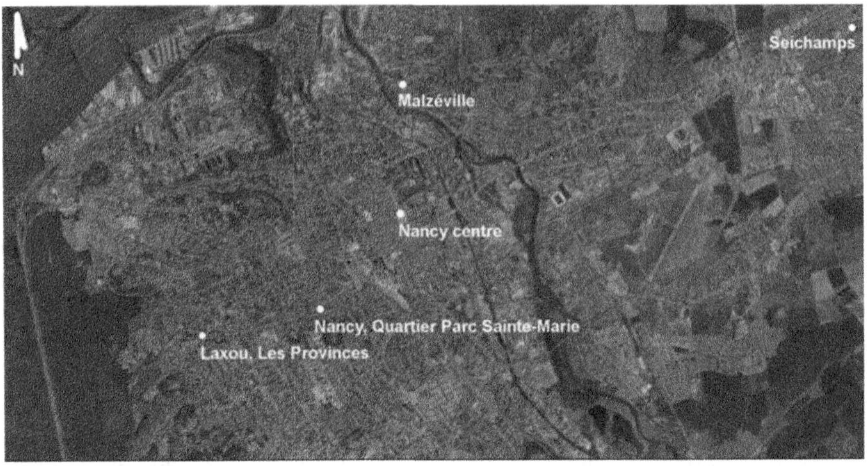

Figure 9.1. *Situation map for the various project sectors, within the Nancy agglomeration (source: geoportail.fr database)*

9.5.2. *Appropriation of the exercise by the students*

9.5.2.1. *Les Provinces*

The first workshop took place in the large *Provinces* area which saw the construction of around 1,000 homes in Laxou between 1955 and 1963. After an initial intervention in the 1990s, the district has experienced a new urban renewal operation as of 2018, as part of the New Urban Renovation Program (NPNRU, *Nouveau programme de rénovation urbaine*). The students arrived on the site shortly after the beginning of the study phase.

In this context, the students made the choice to tackle important and urgent issues. To begin with, it was the issue of the urban living environment's quality that was raised, signaling the need to be able to find proximity, even complicity, with nature (in the broad sense of the term, all living things and natural conditions). As for its continuity, the site's orientation as a support for training in agricultural production has been proposed, particularly with the goal of changing consciousness and creating another imaginary through the local reappropriation of essential know-how, such as truck farming. Finally, the students addressed the issue of thermal renovation in large collective housing units, always in relation to the district's urban transformation.

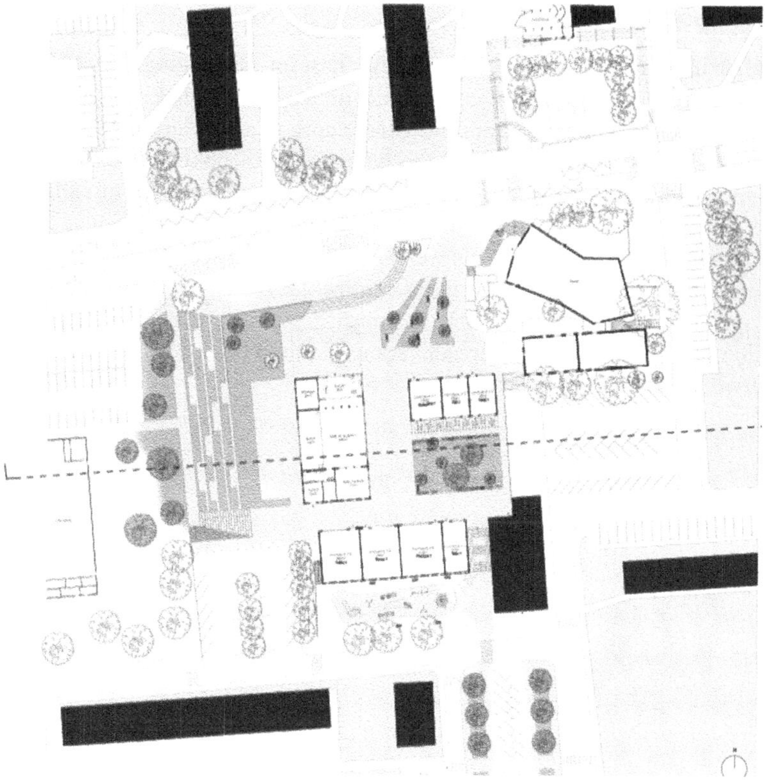

Figure 9.2. *"Local shops", Léa Cordone, student at ENSA Nancy, 2016*

Figure 9.3. *"Picardie Gardens", Arthur Thomas, student at ENSA Nancy, 2016*

9.5.2.2. *Malzéville*

Malzéville, a town bordering Nancy, has an interesting location between Meurthe and the hills, where there are still orchards and truck farming areas. It is made up of different fabrics, from the village core to a few large collective housing operations. In this context, with all the potential, students chose to work on three topics in particular. First of all, they considered the consequences of ending the use of cars. Then, they reflected on the reinvestment of the spaces consequently left empty (parking lots, roads, etc.), as well as alternative forms of mobility. Later, they addressed the training of children in environmental issues, relying on the territory's landscape resources. Finally, truck farming areas gave rise to reflection on the limitation of urbanization by the reactivation of a food belt.

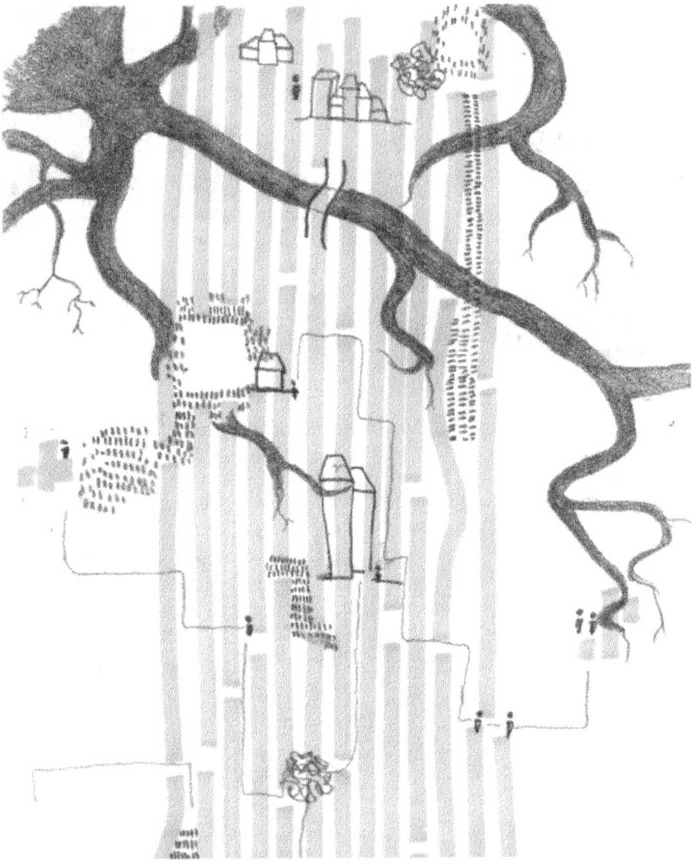

Figure 9.4. *"Educational cheese dairy", Bérénice Plantin-Emeriau, student at ENSA Nancy, 2017*

Figure 9.5. *"Educational trails", Camille Zinutti, student at ENSA Nancy, 2017*

9.5.2.3. *Seichamps*

The third project workshop took place in the context of Seichamps, a city undergoing a demographic boom at the time of the emergence of peri-urban development in Nancy: in addition to a small village core, it is made up of a mostly suburban fabric. The particular characteristic of this workshop is that it was conducted with the collaboration of the *Laboratoire Lorrain de Sciences Sociales*[12],

12 2L2S, University of Lorraine.

which at that time was carrying out research on suburban lifestyles. Two sociologists intervened alongside the students on several occasions during the semester. The area presented yet another peculiar trait: on listening to the inhabitants, they did not seem to come across any problems. How then can we question an area through the prism of ecology when this does not seem to raise any specific problem? The students mainly explored the evolution of this architectural and urban typology through the prism of thermal renovation and the typological evolution of housing and its urban fabric, by calling into question the implications of a gradual abandonment of certain sectors in the context of urban decline.

Figure 9.6. *"Plural foster homes", Camille Hiolin, student at ENSA Nancy, 2019*

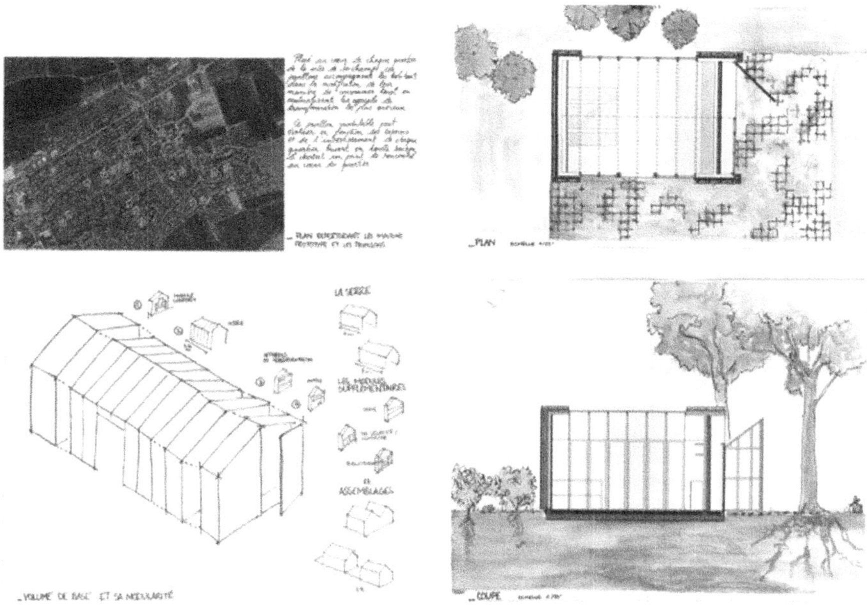

Figure 9.7. *Locally produced food "processing pavilions", Luce Lachaize, ENSA Nancy student, 2019*

9.5.2.4. *The Nancy urban fabric*

The last site explored is that of a classic Nancy urban fragment, namely closed units made up of small three-story buildings on the street front, built at the turn of the 20th century to the west of the historic center, beyond of the boundaries of the railway line. The students were particularly involved in the issue of adapting this fabric – to mineral public spaces but in the heart of a green island – to climate change, and in particular to heat spikes. Sainte-Marie Park, one of the green lungs of the city located just south of the units considered, was integrated into the reflections on urban climate management, as well as on the biological continuity which could be envisaged from the core of the small island.

In this rather rigid fabric, mainly housing private dwellings, students genuinely wondered how the architect could actually intervene, and what the limitations of such an intervention might be. These reflections fully echo the profession as it is practiced today, something which will undoubtedly lead to the other types of construction requests or access to such requests, in order to support populations facing environmental changes that go beyond traditional architecture practice.

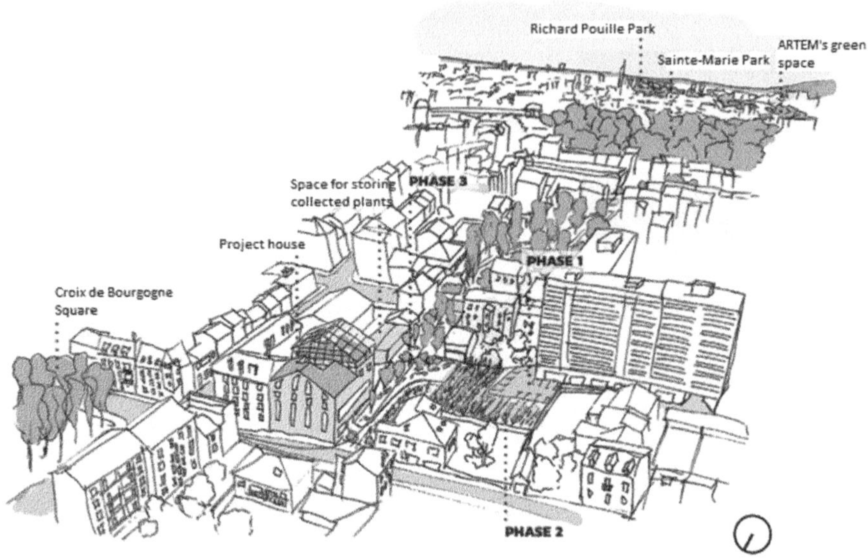

Figure 9.8. *Cloé Barlet, student at ENSA Nancy, 2019*

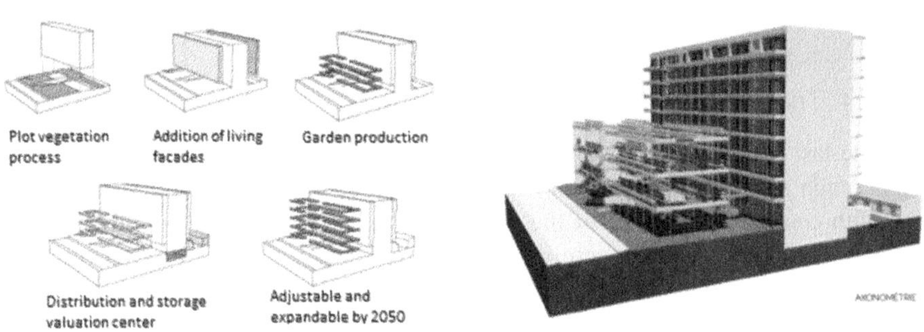

Figure 9.9. *"The living tower, productive and social renewal of a large set", Clara Marquet, student at ENSA Nancy, 2019*

9.6. The Mutations final thesis

The final thesis proposes to take the workshop a step beyond the first semester of Master 2: for the student it involves a role-playing exercise in which they have to picture themselves implementing the most suitable action just as a professional architect would do. While the workshop's operation remains identical to that of the first semester (with a follow-up by all the team members and an increase in autonomous working time), the site is no longer imposed: the student chooses the data for the project, namely the place, the subject and the intervention to be carried out. Leaving this important margin of freedom to the student[13] worked as a convincing first step towards a professional orientation/position.

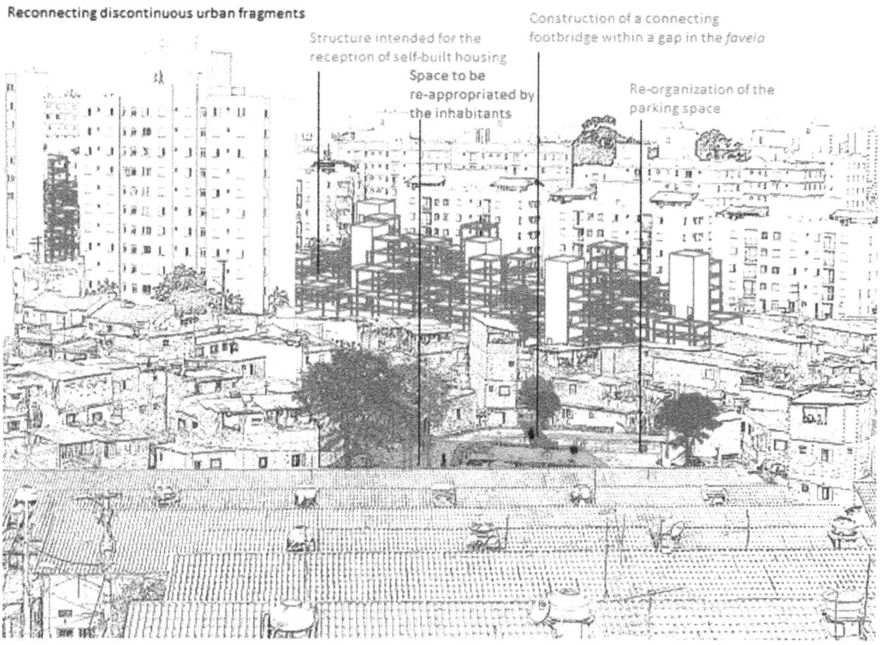

Figure 9.10. *"The Mutirão Jardim Celeste in São Paulo", setting up collective dynamics and self-construction processes to question the relationships between spatial and social integration, Manon Lefranc, student at ENSA Nancy, 2018*

Here is an example of some of the topics already discussed in the framework of the final thesis. The question of the territory brought together many themes, which dealt with reactivation and engagement of local sectors (wood and stone), the implementation of new agricultural logics and territorial development from new modes of public transport. Each time, these points of entry made it possible to question the overall functioning of the territory as well as to assess architectural proposals for activating and/or perpetuating territorial evolution. Secondly, many

13 While the problem is to be assessed by the student, i.e. developed in the most suitable way in accordance with the diagnosis performed, the level and type of intervention at the site (architectural or urban, large or small scale project, etc.) varies from one student to another. Competence mastery expected from an architect will be assessed in this context, the relevance of which will also be taken into account in relation to the various diagnostic elements provided by the student.

topics have developed into a social problem, aiming to apprehend the lifestyles of specific audiences in a different manner. From an alternative school to a living unit for patients with Alzheimer's disease, including the implementation of intergenerational housing, students were keen to reach a variety of audiences, by offering them living spaces enhancing their well-being. Finally, ecological issues were also addressed through the prism of material recovery and gray energy management.

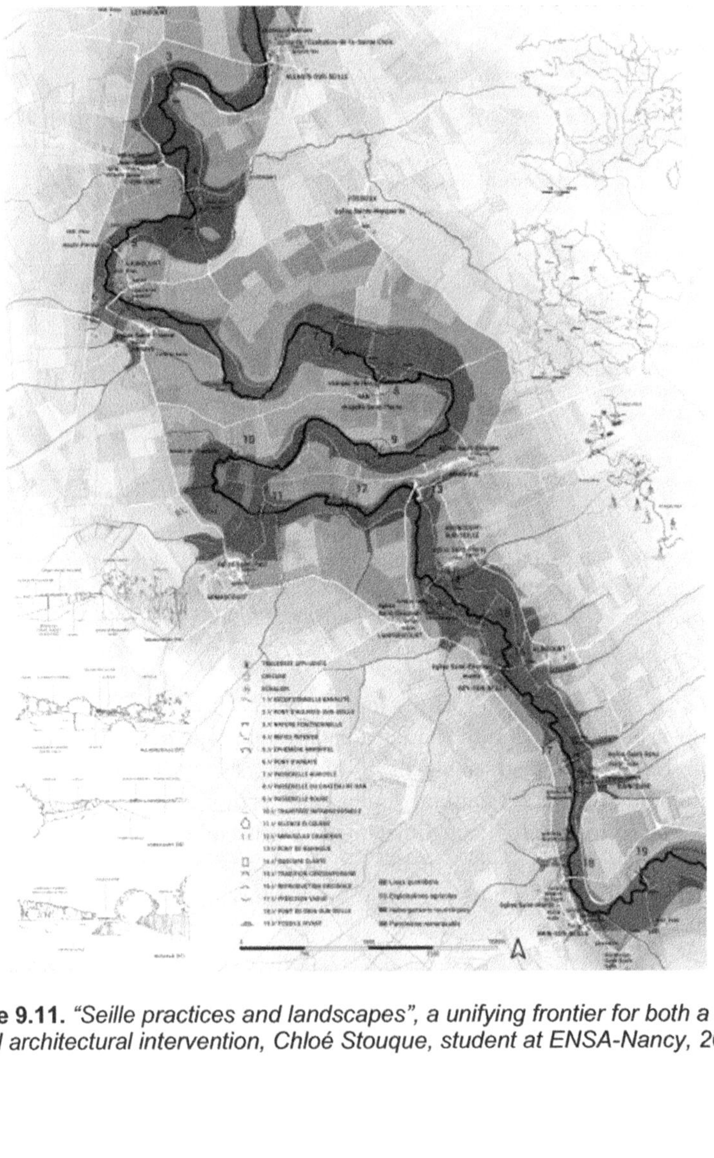

Figure 9.11. *"Seille practices and landscapes", a unifying frontier for both a social and architectural intervention, Chloé Stouque, student at ENSA-Nancy, 2018*

All things considered, the set of proposals mentioned reflects a major difficulty: is it possible – and if so, in what way – to deeply question our ways of thinking and our ways of acting to effectively reduce our environmental impact and adapt effectively to a degraded ecological context? Despite the educational context offered (by means of workshops and RIM meetings, in particular), which tries to create a breeding ground for the emergence of alternative architectural stances, the difficulty in escaping from our cognitive reflexes seems almost untouched… It is therefore essential to continue to work in this direction.

It has been four years since the Mutations team was formed. Our initial reason for existing has not disappeared; on the contrary, it seems more important than ever to continue the work and reflections we have engaged in. The projects proposed by the students and their commitment invite us to do so. Nevertheless, the space, the time, the methods necessary for the emergence of knowledge and know-how geared towards the preservation of the living and the maintenance of lasting relationships within the biosphere have continued to be rethought since the beginning of our work. The RIM meetings as well as the project workshops will be continued, undoubtedly alongside other forms of teaching and research practices, which today continue to fuel collective reflection.

9.7. References

Bird Rose, D. (2019). *Vers des humanités écologiques.* Wildproject, Marseille.

Choné, A., Hajek, I., Hamman, P. (2016). *Guide des humanités environnementales.* Presses Universitaires du Septentrion, Villeneuve d'Ascq.

Curien, É. (2018). *Gion A. Caminada : s'approcher au plus près des choses.* Actes Sud, Arles.

Curien, É. and Fries-Paiola, C. (2018). *Changes 2016–2018.* ENSA Nancy, Nancy.

Fries-Paiola, C. (2017). Géobiologie, fengshui et démarche environnementale dans l'habitat : pratiques professionnelles et représentations. PhD Thesis, University of Lorraine, Lorraine.

Grandcolas, P. (2017). Changement climatique et crise de la biodiversité : la dangereuse alliance. *The Conversation* [Online]. Available at: https://theconversation.com/changement-climatique-et-crise-de-la-biodiversite-la-dangereuse-alliance-83825.

Pelissolo, A. (2019). L'éco-anxiété nous guette, et ce n'est pas forcément une mauvaise nouvelle. *The Conversation* [Online]. Available at: https://theconversation.com/leco-anxiete-nous-guette-et-ce-nest-pas-forcement-une-mauvaise-nouvelle-123028.

Wieviorka, M. (2017). Rapport sur les inégalités : et après ?. *The Conversation* [Online]. Available at: https://theconversation.com/rapport-sur-les-inegalites-et-apres-89318.

Part 3

Teaching and Research on Ecological Transition: From Theory to Practice

10

Form Follows Partnership

10.1. Introduction

Does form follow partnership? Is a partnership of pro-ecology stakeholders a condition to initiate an ecological architecture project? Without claiming academic rigor, this cumulated experience throughout a professional architectural practice between 2015 and 2020 offers a design method based on global and systemic analysis. Starting with a critique of globalization and metropolization, of the "fossil-like" characteristics of current cities, the method transposes systemic analysis to local strategic action for urban ecology. It entails a four-stage process: analyzing the material sectors of the city; re-examining these sectors, thereby re-examining their architectural brief; inventing briefing for inventing architectures; a demonstrator project that illustrates systemic thought. This method highlights the critical importance of stakeholders' partnership when carrying out construction and the power of this partnership to trigger green projects. Finally, what is the role of the architect in this partnership? The architect, whose role has recently deteriorated from creator to service provider, can once again play a central role as a partner at the service of ecological architecture.

Originally entitled "Urban logistics buildings as a reduction in city emissions", this intervention in the "Professional practices"[1] round table intended to respond

---

Chapter written by Damien ANTONI.

1 "Workshop 3: Professionals" whose speakers were: SYVIL experience, Damien ANTONI (ENSAP Lille); Using real calls for competition in architecture workshops: Feedback from the *École Polytechnique de Bruxelles*; Jean-Sébastien DELVIGNE, Séverine HERMAND and Samia BEN RAJEB, (*École Polytechnique de Bruxelles*); Building professionals facing energy challenges, Géraldine MOLINA, Marjorie MUSY, Margot LEFRANC (ENSA Nantes); Games of Adobe, Zoé TRIC and Gian Franco NORIEGA (Amàco).

to two distinct thematic entries: a way of mitigating global warming and providing testimony on one or more projects in progress. The debate that followed the intervention went beyond this simple framework: it focused on the operational conditions for implementing ecological architecture, its barriers and opportunities. Here, a deliberately broader vision of the contribution of the round table to ecological thought in architecture will be exposed. It is based on a perspective of the practice of operational architectural and urban planning projects from the SYVIL architecture workshop, not on an academic research practice. It explores the hypothesis according to which ecology can only be implemented through the application of systemic thought on material flows and on the immaterial conditions which result in the fabrication of architecture and landscape. Repositioning the architect's role as an essential partner of the ecological territory is equivalent to affirming that only a partnership of actors in favor of ecological concerns will make it possible to enforce a truly ecological architectural project.

10.2. The need for systemic thought: think global, act local

Ecology is a problem that affects the entire planet, but whose causes and consequences are local. Nowadays, in the field of architectural practice, the spirit of action "think global, act local" reestablishes a bond with the modernist tradition, according to which, to practice architecture is to have a world vision; it involves knowing the existing conditions of production where the construction of a building is to take place. This is less a question of a *Weltanschauung*[2] – a totalizing conception of a certain idea of the world – than concrete and critical analysis, which is not distant from Marxist analysis, insofar as the material planetary conditions of production are concerned.

In the same way that certain resources are overexploited (sand, etc.) building consumes too much fossil energy (clinker for cement, etc.), and it is therefore fundamental to carry out a systemic analysis of the reasons for which architectures and cities are built. One of these is economic growth; another reason is globalization, which fuels metropolization. Without analyzing what could lay the foundations for degrowth architecture, a task we will leave for others to tackle, it is important to understand the phenomenon of metropolization which has been at work for 60 years now. In point of fact, the economic and demographic growth of

2 Kant defines *Weltanschauung* as "the intuition of a whole [...] which recognizes no other measure, definite, valid for everyone, and invariable, than the absolute whole" (Kant 1914).

metropolises, in France for example, has accentuated the demand for construction to the detriment of medium-sized towns, which have been partly emptied of their inhabitants.

Thus, we can consider that it is illusory to build an ecological architecture *per se*, while the phenomenon generating it is not ecological in itself. In other words, it is essential to think at an urban scale, even a territorial one, to refrain from acting solely at the architectural scale. From the methodological point of view, we approach the "surveying" method developed by Patrick Geddes (2010, p. 162), which he recommended as the first stage of any urban planning method: any urban study, in fact, any architectural project should start with a broad and open understanding of a territory in all of its aspects: economic, social, political, historical, morphological, etc.

10.2.1. *Metropolization as inseparable from globalization*

As Dablanc and Fremont (2015, p. 309) have illustrated in the case of goods transportation, the phenomenon of globalization is fueling the phenomenon of metropolises. In return, metropolises reinforce globalization in that they concentrate economies, flows and capital which are favorable to globalization. In material terms, metropolises are dependent on gigantic material flows of resources to supply themselves, feed themselves, and function on a daily basis. For example, Paris only has three days of food reserves (Basset 2012, p. 128). Metropolises like Panama have been created from nothing thanks to the globalization of goods exchange, namely through international maritime trade.

10.2.2. *"Globalization does not exist. Only globalized sectors exist"*

During a conference at ENSA Paris-Belleville in 2005, Saskia Sassen, one of the theorists of globalization, communicated to the students that "globalization" was simply the sum of interconnected "globalized sectors". This analysis already carried the seeds for change: in order to begin to take back the outrageous globalization of trade and excessive resource consumption, it was necessary to work sector by sector. What followed proved her right, in particular with the public success of the idea of a short circuit, identifying a sector and re-examining it in its deployment in space.

A project methodology thus emerges, using the analysis of sectors as a triggering point to identify relevant programs that will materialize in new architectures and landscapes.

10.2.2.1. *First stage: analyzing the city's material sectors*

In order to avoid tackling a gigantic problem which is difficult to define, it is possible to go sector by sector. "Material sector" stands for a set of companies and communities which constitute a network and make sure that all the phases of a transformation process or material movement are met. Regarding the city, we have identified seven material sectors: raw materials, water, waste, energy, manufacturing, logistics, and food. They also make it possible to draw up an inventory of buildings which are the constructed manifestations of such sectors: logistics platforms, waste reception centers, electrical transformer stations, thermal power stations, etc. Understanding these constructions and acknowledging their existence also means imagining what they could eventually become.

10.2.2.2. *Second stage: re-examining sectors, re-examining their programming*

In the next sequence, these sectors are closely re-examined: which sectors are predatory for the natural environment? The characteristics of these "fossil" sectors have thus been identified (Antoni 2019):

– non-looping, that is to say, beginning an extractivist chain and ending up in a polluting chain;

– centralist, that is to say, deliberately organized into large "power plants";

– long, comprising many stages and checkpoints, across long distances, at the cost of a great expenditure of energy;

– relatively impervious towards each other and in relation to the city, that is to say, they operate following a logic which is specific to their sector;

– incompatible with nature, distinct from, even predatory of the natural ecosystem;

– uniform, faced with the specificities of climate, soil, geography, that is to say, homogenizing, whereas diversity is constitutive of the biosphere.

How can we promote new "fertile" sectors? How can we cross sectors so that food waste can become, for example, a source of energy? How does a wastewater treatment plant become a phyto-purification field, a chemical fertilizer plant, a vermicompost farm?

10.2.2.3. *Third stage: inventing programs for conceiving architectures*

This unveiling of the fertile sectors, and their material functioning, carries with it the germ of a new programming of the building's and landscape's functions. It is this programming that architectural and landscape design will come to shape and "exhaust" ("*épuiser*"), according to the expression by Rem Koolhaas. River station markets, "decentralized" energy production sites, "pick-up points" and "reconditioning" areas, storage buildings and "decentralized" distribution, are all programs to study to make cities more ecological and more circular.

10.2.2.4. *Fourth stage: efficiently applying systemic thought to a demonstrator project*

Once this program has been proposed, identified, stabilized (with its actors, as we will see), it is a question of concretely embodying it within buildings and landscapes. Because ultimately, the field of action of the architect and the landscaper will continue to be that of the concrete and localized project.

Even if designers cannot always institute a systemic change at the project's scale, having experienced this quest for circular programming, they are able to set a precedent, and to profit from this development to set up a demonstrator project.

This example, crystal clear for fellow designers, will be taken up by others, brandished as exemplary proof by other actors, communities, architects and landscapers.

10.3. From systemic analysis to strategic action for urban ecology

Nevertheless, re-examining sectors is not done on its own. This is because these sectors are made up of actors. Promoting this or the short-circuit supply chain, moving towards such low-energy logistics, presupposes knowledge of the alternative channels and actors who manufacture, deploy and conduct them.

Indeed, the architect does not have a vocation to build up a sector; its role can be perceived as a catalyst, as an operational force at the service of project implementation. As such, they can advise the company, association and community on the feasibility conditions and the strategy for setting up one or more of the sector's stages. For example, a polluted land remediation company that wishes to use the waterway will have to find land near the river and establish itself in connection with a building and production site, etc. But its installation on land in the metropolis will require the resolution of multiple technical, real-estate, landscape and architectural integration questions.

Thus, the strategy for reprogramming cities in a more ecological way begins with the identification of the actors likely to carry an ecological project, be them owners, building operators or communities. This research, called actor sourcing[3], seeks to identify the best actors capable of carrying out a project, thus making it possible to position the project's values upstream.

10.3.1. *Form follows partnership*

Far from Sullivan's "form follows function", the current functioning of the actors involved in these sectors is often based on the economy, or on the city's financialization[4]. Today's financing methods increasingly tend to predetermine architectural projects to a large extent. Thus, the formula "form follows finance", coined by Carol Willis (1995, p. 224), becomes more relevant than ever. But if we are to nuance this idea, any project, even when financed, can only see the light of day if it obtains political validation from the community in which it is set. A negotiation, a balanced partnership, can then emerge.

Actually, it is a partnership based on the ecological values of actors, which will be able to endow the architecture, the programming and the global strategy of each of the actors with an ecological frame.

Is it not the partnership of actors favorable to ecology that makes an ecological architectural project possible? The role of communities, and in particular, of metropolises, can be devoted to the establishment of ecological criteria, that is to say, the rules for the partnership game.

The calls for projects recently launched in metropolitan areas were life-size tests of this ability of architects to generate multi-stakeholder projects.

Based on our experience at SYVIL, back in 2017, the P4 project, a sort of "micro-food terminal" to deliver fresh goods to Paris, was the winner of the Call for Sustainable Urban Logistics Project launched by the City of Paris and to be delivered in November 2020. It was carried out by Sogaris, the platform in charge of Ile de France's logistics strategy. The building program was co-designed as a result of the contest with Ecolotrans (an operator working on electric and gas transportation by scooter and gas vehicles for the last kilometer). The building has been designed to be dismantled in a modular way, not only due to ecological

3 "Sourcing" is originally a marketing term which refers to the act of finding suppliers or candidates when purchasing or looking for human resources.

4 The subject of the May/June 2012 edition of *Revue Urbanism*, entitled "*La ville financiarisée*".

reasons, but also because it is built on public property and is part of a Temporary Occupation Agreement for the Public Domain (for 12 to 18 years).

Here, the (removable) shape of the building followed that of the partnership between the actors carrying it out, in which case one could say that "shape follows partnership": the will of all the stakeholders to make the building removable, as well as to operate zero emission vehicles. It was the result of the balance between the ecological and non-ecological forces at stake.

Either a specialist in built situations or in those to be built, manager of complex issues, strong in the art of synthesis, seasoned in project mode, provided that they have been trained in this professional path, the architect is well placed to coordinate the landing of these projects onto a specific sector. But what is the new role that the architect is taking in these partnerships?

10.4. From the designer architect to the partner architect

From this perspective, the architect takes on a new role. Less a creator of an object, the architect becomes the creator of a process.

10.4.1. *From the designer architect to the provider architect*

Confined to the role of creator, the idea of a demiurge inherited from the time of the Fine Arts, the architect has lost its aura. The era of the "star-architects" is undoubtedly over.

From being a creative architect, the profession has been relegated to the rank of façade and partition designer, a provider of architectural tasks. The architect is no longer expected to fulfill a so-called "full" mission from the design to the completion of a building, but to perform a series of tasks, breakable at will, to maximize the value created for each mission-segment required by the customer. But this practice disqualifies the architect's role, as much as it removes responsibility from the act of building.

10.4.2. *From the designer architect to the partner architect*

In order to regain a place, the role as a service provider can be left aside, and a partnership can be established with the other actors involved in the project. In other words, the architect can become essential in the act of proposing and building a project; not only through the monopoly of the building permit, but by the recognition of their added (social, economic, etc.) value during all the stages of the project. It is good to take on many additional roles in order to retain the ability to carry out the core business as a designer: the architect can thus become a business provider, a programmer, an intermediary, a consultant, an assistant to owner, or even an investor in money or in services, as long as these roles are duly recognized as such, and as long as this makes it possible to "influence" the architectural and ecological match.

10.4.3. *Replacing the role of the architect as a city strategist*

At the crossroads between the political and economic worlds, knowing the strategic analyses and scientific publications on ecological urban subjects enables the architect to propose a vision for the city, backed up by legitimate and relevant architecture. For this, the architect can use two different methods. On the one hand, like a Trojan horse introducing ecological values into a world not very inclined to respect them (finance, real estate, etc.), the architect can cultivate this approach, in particular by relying on the communities interested in these issues. On the other hand, the architect can also learn to bring together alternative actors capable of carrying out a real ecological project; in this sense, self-produced participatory housing is a good example of the transformation of a project through the transformation of its financing modes.

10.5. Openness: teaching an attitude more than adaptation to the market

It is difficult to foresee whether there is an economic market for this type of role that the architect, sector analyst, project manager, programmer-strategist, project manager from A to Z could adopt. The experience of SYVIL in the productive city segment does not have enough perspective to institute a "circular productive city" sector, since there is specific, standardized know-how for the design of architecture housing projects. This experience is too singular to offer students a cutting-edge field of study.

However, the scarcity of public procurement, the difficulty and precariousness of the exercise, push architects to transform their practices, to explore other fields: project management, consulting in all its forms, even within companies which are not architectural firms. To prevent the gap from widening between practice and academic training, it is important to update the teaching of architecture in light of these new data.

The teaching of architecture could pay closer attention to the documentation of the real situation of the territory, its actors, barriers and opportunities for the advent of more ecological projects. This teaching could broadly open the field of architecture to urban planning and landscape, but above all, foster an open, research-oriented attitude. This critical attitude, open but nourished by reality, will offer each graduate the key to understanding how to adapt to each new professional situation, much more than the professional training for architectural contests does.

10.6. References

Antoni, D. (2019). La ville circulaire : de la ville fossile à la ville fertile. *Revue Sur-Mesure* [Online]. Available at: http://www.revuesurmesure.fr/issues/nouveaux-visages-ville-active/la-ville-circulaire [Accessed February 2, 2020].

Basset, F. (2012). *Vers l'autonomie alimentaire.* Éditions Rue de l'échiquier, Paris.

Dablanc, L. and Frémont, A. (2015). *La métropole logistique, le transport de marchandises et le territoire des grandes villes*. Armand Colin, Paris.

Geddes, P. (2010). *Cities in Evolution: An Introduction to the Town Planning Movement and to the Study of Civics*. General Books, London.

Kant, I. (1914). *Critique of Judgement*, 2nd edition. Translated with Introduction and Notes by J.H. Bernard, Macmillan, London.

Willis, C. (1995). *Form Follows Finance: Skyscrapers and Skylines in New York and Chicago.* Princeton Architectural Press, New York.

11

Tackling Global Warming in the Ardèche Valley: Project Workshop Narrative

11.1. Introduction

The ENSA Lyon Master 1 workshop, "Strategies for the bioregion" was organized in cooperation with a community of municipalities of the Ardèche valley, or "*Vallée de la Ligne*". Drawing on the theoretical framework of the Italian territorialist school, the workshop examines the interrelationships between human settlements and territory. The issue of global warming was addressed in 2019, during the third year of cooperation. Complex to tackle at the scale of a small valley, the teaching was based in particular on the socio-technical scenarios of the IPCC[1], which made it possible to identify the dynamics of change at work in the territory, in the sense of adaptation, mitigation, or amplification of climate change. Debated during a workshop with local stakeholders, these territorialized scenarios also make it possible to understand a territory's capacity to initiate or amplify its transition. As part of these projections, students identified project situations enabling them to question the process at work, as well as identify the resources for mitigating and adapting to change. Thanks to the images produced, the proposals developed at this stage made it possible to make the impact of a change on the identified places more tangible, while demonstrating the capacity of the territory to cope with change. Through their prospective and contextualized dimension, each of the projects made it possible to demonstrate the necessary transversality of the responses to be provided by relying on a precise understanding of the resources of each territory.

Chapter written by François NOWAKOWSKI.

For a color version of all the figures in this chapter, see www.iste.co.uk/aithaddou/transition.zip.

1 IPCC – International Panel on Climate Change.

11.2. Three-year cooperation between ENSAL and the Ardèche valley

The "Strategies for the bioregion" workshop was built in cooperation with the Ardèche valley, *Vallée de la Ligne*, in 2016. After an initial exploratory sequence in the form of an intensive one-day session, three years of semi-annual project workshops focused on imagining possible future scenarios for this territory, as part of the dynamic of ecological transition. These successive years made it possible to tackle different issues from different scales and territorial situations. The first year focused on the town center of Largentière and its connections with the valley. In the second year, four villages along the valley were studied, offering different configurations and situations. During this second year, the question of living together and mutualization was central to the reflections carried out on each of the villages. The third year made it possible to study a panel of project situations across the valley, through the prism of climate change. This chapter will focus on the third year to retrace the way in which the issue of climate change has been tackled, given its complexity and abstraction, while calling into question the modalities in which these issues have been debated and discussed among the territorial stakeholders.

Figure 11.1. *An educational approach, getting in touch with the territory (photo: Bernadette Forest, 2019)*

11.3. Putting the bioregion to the test

In concrete terms, the "Strategies for the bioregion" workshop offers design situations that enable students to get in touch with a territory and its stakeholders, residents, users and citizens. This strong commitment within and towards a territory echoes the theoretical framework of the workshop, with research inspired by the Italian territorialist school. These works, mainly embodied in the writings of Alberto Magnaghi (2012), propose to rethink the relationships of human societies with the territories in which they are deployed. Evidencing a "territorial crisis" (Magnaghi 2003), the territorialist school proposes to bring to light, to reactivate, processes of co-evolution between the territory's resources and human populations, through an awareness of "territorial inheritance" (Poli 2018). This theoretical framework makes it possible to make ecological sensitivity a more tangible subject, uniting the teaching team around a common topic and representing a shared basis for constant debate with the students involved in the workshop. Indeed, it is not a question of seizing the thought of the territorialist school and mechanically transposing it to other scenarios but, on the contrary, of resonating on its questions and confronting them with other territorial situations. Since the project was not accurately constructed during the first year, the workshop immediately articulated a projectual and theoretical goal, trying to mobilize projects for the benefit of a reflection on territorial resources.

During the four-year cooperation[2] and the three occasions of the workshop, the partner territory was a community of municipalities of *Val-de-Ligne*, located in Ardèche. This intermunicipal association, whose headquarters are located in the small town of Largentière, develops around the *Val-de-Ligne*, named after the river that crosses it. Made up of 11 municipalities and 6,200 inhabitants, it is a very small territory, which narrowly escaped the inter-municipal merger imposed in 2015–16. This was mainly due to its territorial coherence around the same valley, something which enabled elected representatives to highlight the relevance of its perimeter. This rare match between a geographical entity, a valley, and an administrative perimeter was of particular interest for us and represents one of the pillars of our reflection. Two other partners, the Parc naturel régional des Monts d'Ardèche and the *Conseil d'Architecture d'Urbanisme et de l'Environnement* (CAUE) *de l'Ardèche* also supported the process during the three-year workshop.

11.4. Climate change and its consequences: a shared issue?

Almost incidentally, the discussions prior to the third year brought up a shared theme between the teaching team and the territory's elected representatives: climate

---

2 First contacts in 2016 to 2019, the last year when the workshop took place.

change and its impact on methods for building shelters, cultivating the land and working in the valley in the coming decades. For the *Vallée de la Ligne*, and more generally the south of the Ardèche, the summer of 2018 was characterized by a very severe drought which had a significant impact on local agriculture, already in a fragile state. This subject became evident both in the local and the global news thanks to the emergence of "climate strikes" and "climate marches" in the activist field and in the consciousness of younger generations. The theoretical framework of the workshop made it possible to tackle this question from a broader perspective, starting from the territory but trying to understand the social, economic and political dynamics of climate change. This openness was the result of an explicit positioning of the teaching team, confronting the students with the complexity of these issues and opening the architectural and urban debate beyond architectural resolution alone, by calling into question the various frameworks that shape it. This positioning involved making adjustments to the syllabus, questioning the role of users and user expertise, project leaders, the processes and transformations induced by the planned project, etc. This openness also relied on the diversity of the student profiles, brought together in the workshop. Apart from a majority of architecture students, the workshop also welcomed students having enrolled in the "recycling" course of the "City and Urban Environment" Master's degree (VEU, *Ville et environnement urbain*), whose previous training courses were very diverse in nature. This interdisciplinarity constituted an essential resource for broadening architectural questions and putting them into perspective through other reading prisms.

11.5. Climate: a constant that is no longer so?

The issue of climate change and the challenges it poses are difficult to tackle at the scale of a small territory such as *Vallée de la Ligne*. Indeed, while climate was once a constant that influenced human settlements, rapid climate change is shaking up this stability. How can we position dwellings in relation to the sun or the prevailing winds while climatic phenomena are changing the way in which we can benefit from them? The question of permanence, which can be read in vernacular territorial structures, was key for the reflection carried out concerning the impact of climate change in *Vallée de la Ligne*: one of the main challenges of the prospective approach was to question its permanence, as we inherit it, in order to verify sustainability through the hypotheses of change in the valley's climate. This apparently simple question comes up against the difficulty of having climate change models at the valley's level. Relying on their scientists, the international work by the Intergovernmental Panel on Climate Change

(IPCC)[3] developed models at a global scale for each country, in order for these to be transposed and territorialized. But this work does not enable a direct approach to the scale of inhabited spaces and is difficult to mobilize within the framework of a territorial prospective approach considering that snowball effects (such as processes resulting from cross-interactions between different changes (Le Treut 2009)) are not modeled with certainty, and are all the more complex to consider as we move towards a more restricted territorial scale. Aware of the future changes that are to take place, the nature of these uncertain changes has been mobilized, as shown by the new uncertainties they have introduced into the permanence which having guided territorial development until now. More than a set of stable and precise data, it was necessary to collectively establish an array of possibilities according to which interventions would be developed, making it possible to take part and to question the permanence of human settlements undermined by change.

To address the challenges of climate change at the valley's level, the workshop's approach was broken down into three stages, so as to progressively identify the changes at work and define courses of action, taking uncertainties into account.

A first stage of territorial representation was carried out by means of block diagrams, illustrating its social, geophysical and climatic dynamics; these formed the basis for the exchanges with local stakeholders (elected representatives, Chamber of Agriculture, the CAUE, etc.) and residents, in the form of public workshops.

A second stage was based on scenarios inspired by the "socio-economic scenarios" established by the IPCC. The goal at this stage was to adapt the global scenarios to the valley's level, in order to understand not only their potential but also the dynamics currently at work; and with this information, to formulate the challenges together with the elected representatives and territorial stakeholders.

A third stage involving the identification of project situations made it possible to explore response proposals for these challenges. This last stage ended with an exhibition and public workshops to debate proposals.

From this general development, we will retain three questions underlying the workshop's process in order to provide reflective and critical assessment from the educational point of view and following the reception of proposals in the territory. Through three successive angles, we will endeavor to analyze the way in which the educational systems put in place make it possible to efficiently address certain questions and enable students to work with them. In addition, the reception of

---

3 Intergovernmental body created in 1988, which has published reports on climate change since 1990. The last report on which the workshop was based was published in 2014. The next global report will be published in 2022.

projects will also be discussed and analyzed through the various exchange mechanisms put in place and the feedback obtained from the territory stakeholders.

11.6. Territorializing global challenges

The climate is an "object" whose complexity is difficult to grasp: What is it? How is it transformed? How is it possible to adapt the global graphs describing an increase in average land temperatures, or the maps published at the scale of a continent, to a small valley? The work of climatologists is built around models (Le Treut 2009), which integrate a significant number of parameters that interact with one another. These parameters are fed into the models, which are updated and become more complex over the years by integrating more and more data. In the very short time of a semester and with the tools available at a school of architecture, it is impossible to deal with these questions head-on. A first basis for understanding is to distinguish meteorological phenomena from climate, the former being only a manifestation located in time and in space of the major climatic dynamics. Thus, a particularly cold winter is not the sign of attenuation in the warming process, but simply a located manifestation, which can be related to multiple parameters included in a global climate dynamic characterized by seasonal variations. The notion of climate change thus makes it possible to introduce the idea that global warming will have consequences which will reveal their specificity locally, depending on the evolution of climates.

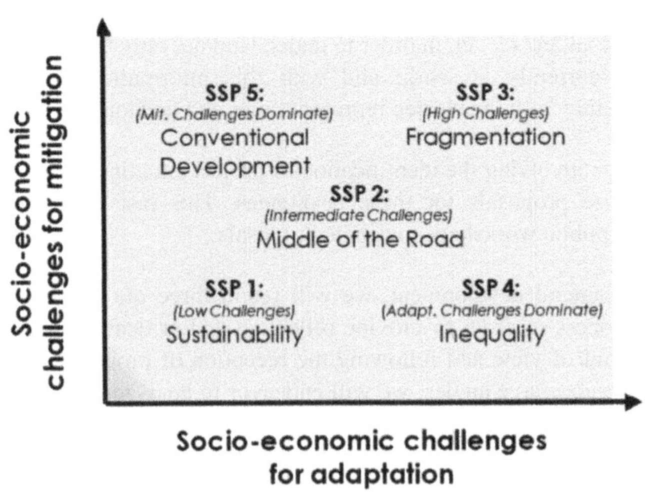

Figure 11.2. *Presentation of the five socio-economic scenarios (source: Nigel* et al. *2011)*

To take a step aside and avoid focusing on the environmental dimension of climate change alone, the workshop took up the socio-economic scenarios (Shared Socioeconomic Pathways, SSP) established by the IPCC in its 5th report, published in 2014[4]. These scenarios compare the trajectory of warming (+2°C, +4°C, +6°C, etc.) with economic, social, political and technical developments. Among the five scenarios developed by the IPCC experts, three (scenarios 1, 4 and 5), proposing highly contrasting trajectories, were the subject of territorialization within the workshop's framework.

These general and global scenarios make it possible to move from the relative abstraction of climate to transformations which can be transposed to a territory. Thus, the first scenario (SSP1), which proposes a rapid transition to a society oriented towards greater equality, emphasizes "mutualization" and sharing, the end of fossil fuels and other production methods which destroy the biosphere and consume large amounts of energy or raw materials, in view of detecting the dynamics currently visible in the territory of *Vallée de Ligne* which go in this direction. The forms of alter-mobility, the renewal (still on very small scales) of certain forms of agriculture based on vernacular know-how, the initiatives of inhabitants for sharing or pooling resources can constitute the basis for a more global transformation as part of this transition trajectory. There are obviously many drawbacks, since this scenario, which goes without saying, is more or less conditioned by a severe questioning of the capitalist system. This first scenario is part of a global trajectory of strong climate change mitigation to reduce adaptation needs, since global warming could be contained below 2°C. In this scenario, as nuanced on the scale of the *Vallée de la Ligne*, the valley's existing villages are re-occupied by populations committed to the dynamics of short circuits and frugal and local development based on recycling and sharing. The places which make it possible to develop this common awareness (squares, gardens, and village outskirts) take on greater importance.

The second scenario explored (SSP4) starts from an opposite hypothesis: as inequalities increase, the wealthiest populations can invest on adaptation, whereas the poorest populations have to face significant deterioration in their quality of life. This scenario brings to light the social inequalities and economic precariousness of a part of the population of the Ardèche territory, and in particular of the *Vallée de la Ligne*. The territory, which is already functioning in a fragmented manner due to the dynamics of demographic decline, is witnessing an increase in its appeal as a secondary residence setting for wealthy populations.

4 The IPCC reports are available on its website: IPCC.ch. The scenarios discussed in this chapter (prefixed with "SSP") are available at the following address: https://www.carbonbrief.org/explainer-how-shared-socioeconomic-pathways-explore-future-climate-change.

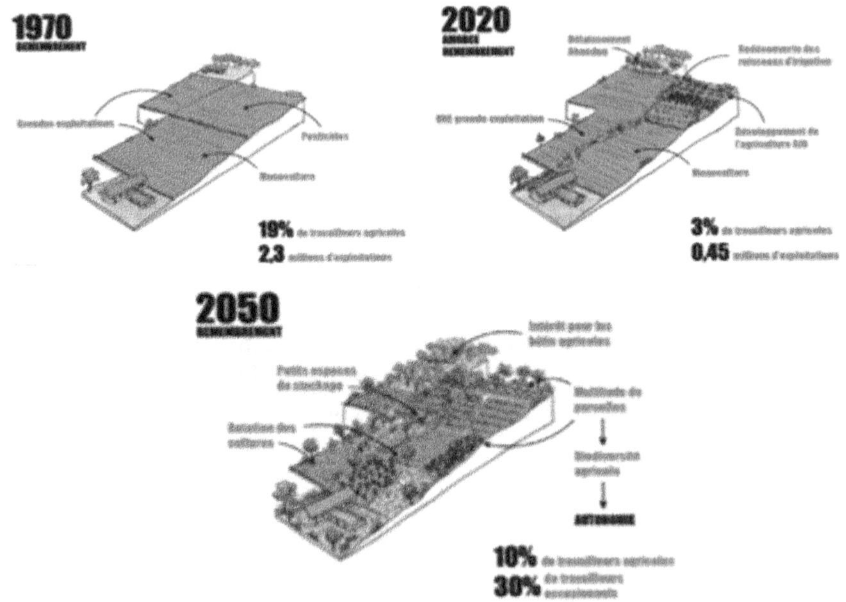

Figure 11.3. *An agricultural transition process imagined from the IPCC, SSP 1 scenario (source: Laetitia Fraichard, Walid Zrari, Yinbo Dai, Johanna Pinherio-Rodrigues, Rudy Fieremonte, Willy Mourey, Chloé Bour, Amandine Marti-Nafti, Romane Petit)*

The third scenario (SSP5) is based on a hypothesis of strong technological development making it possible to reinforce adaptation efforts without real mitigation results, though. The students focused their local version of this scenario on the question of tourism, which clearly illustrates some of the contradictions in the territory's development for the past half-century. The tourist season, which is characterized by an influx of people from May to September, and the accompanying reopening of many shops, services and accommodation facilities, is also the period of the year when shortages of certain resources becomes more problematic. During the summer season, the territory is in a situation that verges on water-stress (drought), as the competition between various needs (agriculture, human needs) is strong and reinforced by a seasonal increase in the population. Due to resource scarcity, this scenario promises a collapse of this tourism model within one generation.

Figure 11.4. *Images to fuel the debate on the impact of tourism on the territory (source: Mallaury Abadie, Estelle Cossais, Emma Duplanil, Diego Fuentes-Santos, Sarah Ielissof, Anne Loustalet, Louise Ranc, Sarah Rener, Aude Takasi)*

Each of these scenarios reveals issues that affect the territory at present, and makes it possible to extrapolate the consequences of the transformations which can already be foreseen. During the time of exchange which followed the presentation of the scenarios[5], territorial stakeholders testified to a highly "true-to-life" character, despite the fairly free extrapolations that were made, the situations observed and deciphered. To formulate the challenges, it seemed difficult to deviate from the frameworks adopted by current territorial policies, for example the competition between territories. As such, scenario 2, postulating an increase in social and environmental inequalities, may have appeared interesting for some elected representatives, since it was based on the particular appeal of the territory to wealthy populations. Scenario 1, *a priori* the most desirable one, since it makes it possible to achieve a trajectory of sustainable global warming mitigation, seems very close to the dynamics currently visible through the multiple initiatives characterizing the Ardèche region, but also very distant, as placing multiple alternatives within a "system" still seems remote. At a very small scale, this illustrates certain

5 During a feedback and discussion session, March 29, 2019.

discrepancies between citizens and appointed representatives in how they anticipate and envision changes.

11.7. Associating experiences with broader dynamics

Through the various opportunities for exchange and the presentation of their reflections, the workshop approach voluntarily put the students in direct contact with the territory and those living in it, imagining it, experiencing it... This educational choice responds to the challenge of training future professionals in spatial and territorial transformation to be able to have fruitful "exchanges" with all those who, despite being non-architects, will nevertheless be in direct relation with future designers. This skill is not innate and is not related to certain "know-how", but stems from a performing capability: formulating questions, generating the appropriate support for exchanges, developing an ability to listen, to reformulate, to debate. In order to address the issue of climate, as soon as students were immersed in the territory for the first time, it seemed essential to integrate moments of exchange and meeting with the elected representatives and inhabitants, by settling down in four villages to discuss climate change and the way in which it is perceived at present. The discussions also intended to provide a better understanding of the vernacular inhabiting practices in relation to the climate: How can we take the wind into account? The sun? The rain?

Figure 11.5. *Time for discussion and exchanges in the training of future designers (photographs: Bernadette Forest)*

These exchanges immediately revealed the above-mentioned confusion between climate and weather, on which climate skeptics base part of their argument. But the exchanges with the inhabitants showed that global awareness about change is widely shared, even if its impacts still seem difficult to grasp. Among the elements mentioned was the increase in periods of drought, the force of the rain and heatwaves, which can make it difficult to work on certain mineralized spaces. These elements are in direct relation with the designer's know-how and become essential for various projects, since they often relate to very concrete aspects which are part of the local inhabiting practices. The first public workshop was based on the block diagrams of different sections of the valley. This approach quickly proved rather unsuitable for exchanges when it is based on specific situations which cannot necessarily be extrapolated to the situations experienced by the inhabitants and their living area. We thus realized the difficulty of moving from a concrete experience to a more globalizing representation, especially when this representation is very illustrative and based on a real situation. The exchanges therefore were based more on discussion, sometimes even on maps, rather than block diagrams.

Other opportunities for exchange arose during the semester, in a more or less formal way. The semester ended with two days of events of differing formats: treasure hunt, hiking, film debate, exhibition, "voting" for projects according to different categories[6], etc. Moreover, by returning to the territory, the students could gradually build up a certain familiarity with it. When working in a hamlet comprising very few houses or the village's square, it is very difficult not to be spotted and quickly approached by local inhabitants. The three years of presence in the territory also brought recognition and sometimes an expectation on the part of certain inhabitants and local stakeholders. These informal meetings also made it possible to deepen the knowledge of inhabiting practices. In Méjanne, the long tour of a house with its inhabitant helped us to understand the proper management for opening of windows and shutters during the day, taking into account the course of the sun, so as to ventilate the house and avoid overheating.

11.8. The project as a means of testing, demonstrating, convincing

As expressed by the CAUE de l'Ardèche at the start of the three-year cooperation, one of the great strengths of the projects developed at ENSA schools is making certain principles, orientations and territorial transformation more tangible, thanks to the representations brought forward. The accuracy of these representations, which translate the transformations projected, also make it possible to enter into a discussion with the stakeholders and the territory's appointed representatives based on concrete and understandable elements, moving beyond general principles to collectively question ways of taking action. With the global challenge of climate change, the gaps between global awareness and the possibility of acting locally (having a grip and an impact on daily life) seem enormous. Throughout the workshop, the scenarios designed to formulate challenges and to deepen the understanding of the dynamics at work in the territory come after a project phase

6 "The project that makes you dream", "the necessary project to mitigate climate change", "the important project to be carried out".

inspired by the situations resulting from said challenges. This stage required some risk-taking, by choosing a situation whose exploration would have to be pursued in order to support (or not) the intuition that made its identification possible in the first place.

Figure 11.6. *View of the village extension site, before starting building works, in 2018 (photo: Bernadette Forest, 2018)*

In the village of Rocher[7], a project currently being implemented justified the need for a counter-proposal. At its core, the village of Rocher benefits from the presence of a Residential Care Unit (EHPAD, *Établissement d'hébergement pour personnes âgées dépendantes*) which functions as a day hospital and a serviced residence for the elderly. The heart of the town adjoining the old manor which now houses the health complex has gradually expanded in the form of single family houses. A new extension is planned on the land adjacent to the town center. By 2019, this extension had reached stage where electrical boxes had appeared and a temporary pavement structure had been put in place: now, the houses are expected to take shape in the middle of the plots delimited by stakes. The group of students will take up this project to call into question its links with the town center, following the hypothesis that strengthening the town center is likely to favor the emergence of new communal forms and mutualization which can be inscribed in the dynamic of climate change mitigation through the development of local life. Their proposals were broken down into three hypotheses, starting from a simple adjustment of the

7 Project developed by Willy Mourey, Laetitia Fraichard and Evegueniya Moskalyuk.

project being implemented to a more significant development, reusing the technical infrastructure already in place. The first scenario would not modify the structure of the public spaces created but would intervene on their connections with the town center, on the hedges, on the management of rainwater and on parking space sharing, benefiting from an existing right-of-way located on the fringes of the project. The second scenario proposed to intervene on the house's layout: by imposing joint ownership, the use of the plots would be optimized and the thermal quality of the housing, improved. The rule change would make it possible to share access and anticipate densification, hindering the establishment of houses in the middle of the plot. The layout on the plot line and along the street would also make it possible to envisage service activities which could potentially take place on the ground floor. The third scenario, requiring a greater modification of the project underway, would strengthen the alley's network between the town center and the rest of the village through the extension sector, in order to fully integrate it with the town center.

Figure 11.7. *Plan of the three alternative scenarios for the planned suburban extension in the village of Rocher (source: Willy Mourrey, Laetitia Fraichard, Evegueniya Moskalyuk)*

At first glance, these three proposals could seem quite distant from the climate change adaptation and mitigation challenges the workshop proposes to deal with. But actually, they make it possible to tackle changes in lifestyle, in relation to mobility and local resources which residential development projects do not take into account. While in general the principle of residential extension must obviously be called into question in favor of reoccupying existing spaces, the pragmatic approach proposed here shows that a project already under way is not irremediable and can still evolve and offer a better response to contemporary ecological issues. The project also shows that some – *a priori* harmless – challenges, such as slight changes in the size and layout of streets, block size, pedestrian walkways, building layout and plot subdivision can turn a residential project focusing on car travel and individuality into a project promoting common benefits and creating the conditions for the emergence of new lifestyles, thus favoring links with the nearby territory. In this sense, this proposal also makes it possible to consider the evolution of other suburban sectors existing in the territory.

Figure 11.8. *A river forming an oasis of freshness in the heart of the medieval village of Largentière (photo: François Nowakowski, 2017)*

The proposals developed for the town center of Largentière[8] focus on the challenge of making more fresh spaces available for a territory that experiences periods of increasingly intense and regular summer heatwaves. The bed of the Ligne, a river crossing the valley, constitutes a fresh space which, despite the lower flow of the river in summer, features important plant cover that contrasts with the mineral character of the town center and its public spaces. The town center bends close to the river due to its medieval fortifications, built on the river bank. Only three bridges and a few private entrances interrupt the town center's frontage along the Ligne. Following on the two projects developed during the first year of the workshop and envisaging breakthroughs in the town's facade towards the river and

8 Proposals developed by Victor Gerez, Louise Ranc, Sarah Rener, Sybille Thubert.

the establishment of a path in its bed, the project imagined in 2019 also offers additional access to the river and to fresh areas, by working on the buildings bordering it.

This intervention constitutes the starting point for a network of public spaces reintroducing planted areas in the town center in order to provide users and residents with shaded spaces that are protected from the sun. The few planned demolitions will restore livability to certain buildings which do not benefit from any outdoor spaces and very little direct solar gain, due to the alleys' narrowness. Shared gardens on the river banks will also make it possible to reintroduce a small productive activity. These gardening spaces adjoining the new headquarters of the community of municipalities could be set within a historic building, incorporating the now unoccupied old medieval tower by the river. In this second example, we see that the integration of an existing natural "infrastructure", the river and its wooded corridor, makes it possible to reflect on the mitigation of the impact of global warming in a town center, in order to preserve its habitability. Here, the constructed structure of the town center is re-examined in light of the intersecting challenges between contemporary living and climatic issues, which can sometimes be opposed: Could de-densifying the town center building in order to improve the amount of sunlight in the dwellings likely expose them to the effects of heatwaves even more? Could de-densification contribute to the extension of urbanized areas? Without providing a definitive answer, this project has sparked a fruitful discussion on these questions with the stakeholders and the residents.

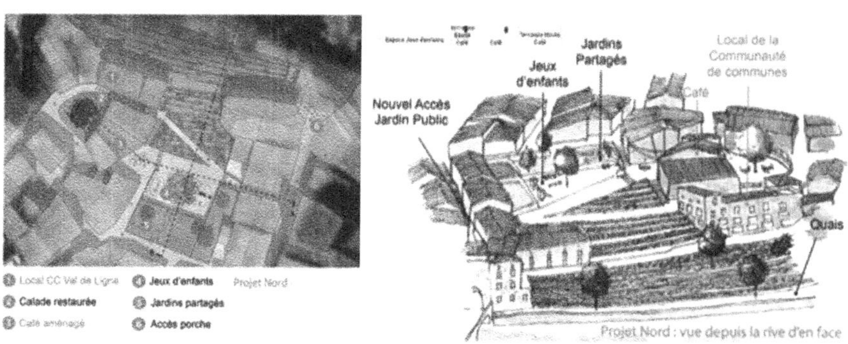

Figure 11.9. *Plan of a transformed island to connect the town center with the Ligne (source: Victor Gerez, Louise Ranc, Sarah Rener, Sybille Thubert)*

The village of Montreal[9] stands out for its position looking down on the *Vallée de la Ligne*, which favored the establishment of a castle intended to control the

9 Project developed by Anne Loustalet, Amandine Marti-Nafti, Romane Petit, Aude Takasi.

valley, as well as to access to the small town of Largentière and its silver mines. The village's appeal, due to its landscaped location, results in diffuse suburban extensions bordering the old agricultural paths connecting farms and silkworm rearing houses with the village and the castle. The first stage of the project envisaged in this municipality consists of developing a pedestrian loop, connecting the various urbanized sectors. This 3-km loop is also intended to make certain resources and vernacular know-how of the territory visible, through the practice of walking and cycling. The cycling rhythm and even more, the walking rhythm makes it possible to re-establish a bond with the place, with the ground, with the landscapes encountered.

Figure 11.10. *Cross-section of a water management and land cultivation device, made visible by the projected pedestrian loop (source: Anne Loustalet, Amandine Marti-Nafti, Romane Petit, Aude Takasi)*

This loop symbolically connects the diffuse suburban areas to the village and to a large open space, located at its center. This public space, with its rugged topography, tells the story of the evolution of the village of Montreal, when it was a very dense hamlet nestled against the castle and its metamorphosis into a village-street, along the road from the *Vallée de Ligne* to other villages located on the plateau. This space, whose public status is mainly based on its different names (*Place du Mûrier, Place du Calvaire*, etc.) is today more of an in-between the different parts of the town, than a real unifying space for the village. The project plans to open up further towards the valley, through a green amphitheater providing access to the history of the village of Montreal and its relationship with the territory. This amphitheater is imagined as a place for debate and democratic deliberation at the scale of the valley, reinterpreting the principles of the agora from the Greek city, whose spatial organization and territorial situation were significant for a particular project of society. By inviting people to rediscover the territory through strolling, it is hoped that visibility will be restored to its vernacular heritage, as well as making it understandable once again.

Figure 11.11. *A large open space in the heart of the village of Montreal (photo: François Nowakowski, 2019)*

Following the spirit of the bioregion, the large public space and deliberation area would stage a territorial project aiming to re-establish a bond using a process of co-evolution with the territory, by providing it with a visible and explicit place in public debate. This project questions the spaces of democracy in a context of ecological transition: What forms, what new practices can make it possible to again find a project of ecological democracy, in touch with the challenges of climate change? Without tackling the issue of climate change directly, this project sets the problem within a broader reflection on the relationship of human settlements with the territory, embodied through places of democratic deliberation and the way in which important decisions are made.

These three examples illustrate the way in which the issues formulated and debated through the scenarios come to be embodied, to be territorialized and take shape in significant places of the territory. The planned interventions are of a different nature, since they can relate to a new narrative, illustrated by the project envisaged in Montreal, as a response to the challenges of adapting to climate change or as an outline for of change mitigation processes. Three other proposals, which we do not develop here, focus on the issue of self-sufficiency at the scale of a hamlet, the evolution of wooden shelters thanks to new uses of chestnut wood, ways of reviving town centers through local agriculture and new places for sharing.

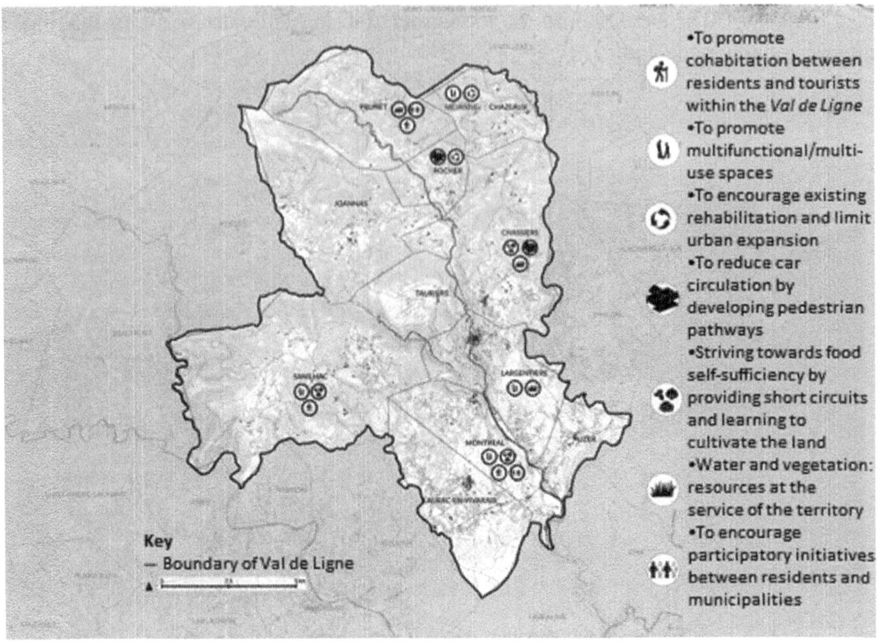

Figure 11.12. *Map showing the location of the different project situations and the challenges addressed (source: ENSAL, Workshop for the bioregion)*

11.9. Projects that open a public debate and lead to the emergence of courses of action

The workshop's brief duration (from February to June), repeated on three occasions, is a very short time compared to the long time taken for the transformation of a territory. The impact of what has been said, exchanged, discussed, projected is always difficult to pinpoint because none of the proposals developed within the framework of the educational workshop is intended to be directly implemented. From this follows that the prospective dimension, shifted in relation to explicitly formulated needs, is a central issue for the workshop: students are invited to consider projections bringing conscious alternatives to certain processes which can help to deterritorialize lifestyles, the way in which human activities reinforce the value and the sense of local resources. Through the complex question of climate change, the scenarios, challenges and projects developed tend to demonstrate that local responses can only be transversal, mobilizing a variety of resources and stakeholders to achieve a transition process. Students are committed to fitting their proposals into the context of ecological transformation whose architectural form is not only a vector – by means of what it shows and what it demonstrates – but also a consequence, recording the transformation of these places into the broader process of territorial evolution. The framework of the educational workshop, based on project formulation, is more than appropriate for reconstituting these links, which other modes of thought and action tend to segment and consider separately. By making possible changes something tangible, the projects contribute to opening a concrete and contextualized discussion on the impact of the various transformations and their interdependencies. This palpable dimension offered by the figurative project constitutes one of the possible contributions of (trainee) architects and various professions related to the design of space and territory (Amboise and Marcel 2015), to a collective awareness of the value of the territory, of these resources and of its potential to be part of a process of social, economic and ecological transition, made all the more necessary by climate change.

11.10. References

Ambroise, R. and Marcel, O. (2015). *Aménager les paysages de l'après-pétrole*. Editions Charles Léopold Mayer, Paris.

Breon, F-M. and Luneau, G. (2018). *Atlas du climat, face au défi du réchauffement*. Autrement, Paris.

Le Treut, H. (2009). *Nouveau climat sur la terre : comprendre, prédire, réagir*. Flammarion, Paris.

Magnaghi, A. (2003). *Le projet local*. Mardaga, Sprimont.

Magnaghi, A. (2012). *La biorégion urbaine, petit traité sur le territoire bien commun*. Eterotopia France/Rhizome, Paris.

Nigel, A., Kram, T., Carter, T., Ebi, K., Edmonds, J., Hallegatte, S., Kriegler, E., Mathur, R., O'Neill, B., Riahi, K., Winkler, H., Van Vuuren, D., Zwickel, T. (2011). A framework for a new generation of socioeconomic scenarios for climate change impact, adaptation, vulnerability and mitigation research [Online]. Available at: http://www.isp.ucar.edu/socio-economic-pathways.

Poli, D. (2018). *Formes et figures du projet local, la patrimonialisation contemporaine du territoire*. Eterotopia, Paris.

12

Learning from Toulouse: Lessons from the Bioclimatic Houses of Jean-Pierre Cordier

12.1. Introduction

The bioclimatic approach in architecture is relatively old and has been known in France at least since the mid-1970s. First developed in the USA, the bioclimatic approach spread quickly in Europe after the first oil crisis of 1973. Jean-Pierre Cordier is an important and influential figure in the bioclimatic approach in France and has contributed to the spread of this design method through its buildings and its teaching. The study of some of his buildings seems to be a source of valuable lessons for us because it can show the originality of the bioclimatic approach for the understanding of a regional climate: the climate of Toulouse. Those buildings also demonstrate the emerging interest in climate-based design and energy savings. This chapter aims to understand the motivations, architectural and constructive choices, but also the errors which appear in various houses built by Jean-Pierre Cordier between 1981 and 1986. Interviews with the inhabitants have been conducted in order to understand the thermal behavior of these houses and their comfort performance. In light of this study, it seems that the role given to the resident and the educational dimension of bioclimatic houses are a source of inspiration for contemporary designers and architects.

Nowadays, the bioclimatic approach in architecture is demanded by certain architects and design enterprises. It conveys the idea of taking climatic constraints

---

Chapter written by Clément GAILLARD.

For a color version of all the figures in this chapter, see www.iste.co.uk/aithaddou/transition.zip.

into consideration during the development of projects, with a view to saving energy and resources. However, the bioclimatic approach has a history of its own and has evolved over time, among other things under the influence of various oil crises and research on alternative energies (solar, wind turbines, etc.). Originally, bioclimatic design was defined as *the integration of climatic and meteorological data so as to solve architectural and construction problems*, and the ideal of comfort should guide the choices adopted[1]. The sources of bioclimatic design date back to the 1950s, but the bioclimatic approach in architecture was really born at the beginning of the 1970s, in the United States. Later, it was developed in France and other European countries throughout the 1970s. In France this period was marked by the translation and publication of important works, such as the manuals by American architects David Wright (1978, p. 245) and Edward Mazria (1979, p. 339), or publications in French on bioclimatic architecture, such as those by Jean-Louis Izard (Izard and Guyot 1979, p. 131). Many figures have marked the bioclimatic trend in France by their achievements between the mid-1970s and the 1980s, as is the case for Michel Gerber[2], Frédéric Nicolas[3] and also Jean-Pierre Cordier. Jean-Pierre Cordier was a Toulouse architect born in 1937 who built many houses in the Toulouse region. He was also the author of important publications on thermal considerations for construction, as well as on the integration of natural energies into dwellings (Cordier 2013, p. 282). Jean-Pierre Cordier's achievements were widely published and shared in the 1980s, making him one of the most important representatives of bioclimatic architecture in France and one of the pioneers of sustainable architecture.

Bioclimatic design is considered to be at the center of the first energy concerns in architecture. While this idea is true in the sense that bioclimatic considerations have foreshadowed a large number of the criteria gradually integrated into thermal regulations, the aim of the bioclimatic approach was mainly the creation of *new forms of regional architecture*, as evidenced by the first manual of bioclimatic architecture published in 1963 by Victor Olgyay and subtitled "Bioclimatic approach to architectural regionalism" (Olgyay 1963, p. 190). After the modern chapter and the globalization of forms, the question was to enhance the forms of climatically adapted architecture and to find a new regionalism in architecture

1 According to Georges and Jeanne-Marie Alexandroff, bioclimatic architecture designates "the optimization of the relationship between the dwelling and the climate, with a view to creating 'comfortable' atmospheres through the use of specifically architectural means" (Alexandroff and Alexandroff 1982, p. 216).

2 Swiss-origin architect, trained in the United States, most of whose work is located in the Pyrénées-Orientales.

3 Architect first associated with Marc Vaye and Jean-Pierre Traisnel, authors of the book *La Face cachée du soleil* ("The Hidden Side of the Sun"), 1974, a true bestseller devoted to the use of solar energy in the home.

meeting bioclimatic requirements. This regionalism was also guided by the desire to reestablish some continuity with vernacular and traditional architecture, insofar as certain vernacular construction techniques appear to be well adapted from a climatic point of view (Gaillard 2020). Challenges arose in relation to energy savings and passive design. However, in his practice, Jean-Pierre Cordier is known to have combined the search for regionalism with the requirements of energy savings necessarily correlated with climatic data. Therefore, studying the achievements of Jean-Pierre Cordier is likely to provide us with two distinct but necessarily complementary lessons: How was the bioclimatic approach implemented in architecture at that moment, and what lessons can we learn today from the achievements of that period? Offering a critical examination of Jean-Pierre Cordier's achievements will enable us to examine these two questions. We therefore hypothesize that these already ancient achievements are a source of valuable lessons for the present time. If this hypothesis is correct, the topicality of the bioclimatic approach and the achievements of this movement for the climatic and ecological challenges of our time should be clearer.

The houses chosen for this study are all located in Toulouse and were built between 1981 and 1986. These houses, sometimes inhabited for almost 40 years, provide us with the precious testimony of their dwellers. This study was carried out as part of my doctoral research which focuses on the birth and development of bioclimatic design in France and the United States[4]. Each house was visited in the company of Jean-Pierre Cordier in order to interview the inhabitants and collect as much information as possible on the achievements and performance of the houses. Toulouse is a city characterized by a temperate climate with fairly distinct periods of dry heat and sometimes strong dry wind. We study these creations by following the constructive and architectural choices of Jean-Pierre Cordier. We approach his choice of volume and exposure step by step, then his research on the storage of solar energy, and we finally examine the evolution of the role of greenhouses in his constructions.

12.2. The volume and exposure of a bioclimatic house

To properly detail a construction requires common sense, but contrary to what Descartes wrote at the beginning of the *Discourse on Method*, common sense is not the most common thing in the world. The individual houses of Jean-Pierre Cordier are a real lesson in building orientation. Often freely installed on large plots, they are perfectly exposed, and this exposure is readable through the clear differentiation of facades. The differentiation of facades is today a crucial point for sustainable design in architecture (Jourda 2012, p. 86). In a bioclimatic house, orientation readability

4 Thesis in preparation at the University of Paris 1 at the Institute of Geography under the supervision of Sabine Barles.

seems as important as orientation itself. The first bioclimatic houses were generally characterized by their orientation and the presence of a greenhouse on south facing facades, making it possible to enhance thermal contributions in winter. The Monchoux family house (Figure 12.1), which was built in 1981, perfectly matches this picture. There is a huge greenhouse on the south facing facade. To the north, the openings are kept to a minimum, whereas the roof drops to the ground (Figure 12.2). The few windows to the east and west on the ground floor are protected by a trellis. The volume of the house is determined by a compactness requirement, making it possible to minimize thermal losses. The hemisphere corresponds to the most interesting compact shape which minimizes exchange surfaces with the exterior, while maximizing interior volume. But considering that the volume in a hemisphere is not the most practical to arrange – unless one is working with a geodesic structure – Jean-Pierre Cordier opted for a compromise: a volume formed by combining an octagon and a truncated pyramid (Figure 12.3). This thermal requirement explains the octagonal base plan for Jean-Pierre Cordier's houses, which can be found in many of his constructions. However, room layout is not facilitated by this plan and raises some issues in terms of use. Despite this, the reduction in thermal losses has been calculated at around 64%, in comparison to a house with the same area. This gain confirms the choice for this type of volume.

Figure 12.1. *South facade of the Monchoux family house (source: author's photograph)*

This search for energy savings through the orientation and openings layout characterizes the movement of so-called "passive" houses in architecture. While the term "passive" building is still used today, it actually finds its origin in the work on solar energy which followed the first oil crisis. The term "passive design" was ubiquitous in solar energy publications, as early as the mid-1970s, when it referred to any system operating on solar energy to produce work or to transform energy, and this, *involving a minimum of mechanical means*. The search for a passive architecture consisted of optimizing the volume and openings' layout according to solar radiation and room layout.

Figure 12.2. *North facade of the Monchoux family house (source: author's photograph)*

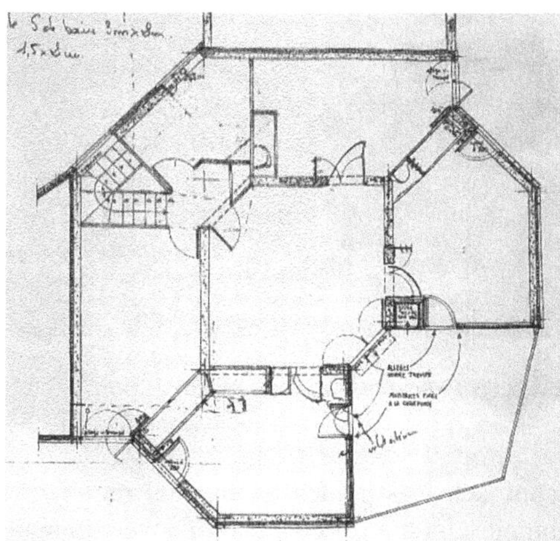

Figure 12.3. *Plan of the Monchoux family house (source: archives of Jean-Pierre Cordier)*

But while choosing a compact volume reduces thermal losses, this reduction also depends on the choice of materials for the building envelope. On this point, Jean-Pierre Cordier broke new ground by proposing to carry out an experiment during the construction of a set of five houses on a housing estate in *Les Pradettes* (Figure 12.4). Built in 1986, these five houses made use of five different construction techniques: the insulation is either internal or external, with air spaces of varying thickness, the supporting structure is built in concrete blocks, in Monomur brick or using a wooden frame. They are all of the same volume and 90% of the glazed surfaces face the southern facade. According to the measurements taken during the first year after construction, the timber frame house with the greatest amount of insulation produced the best energy savings. But for measurements carried out in uninhabited houses, it is still difficult to interpret this result since the internal contributions, those made by the dweller and their equipment could not be taken into account. All things considered, this experimental approach clearly reflects the attitude of the first bioclimatic architects who sought to confirm and to measure their hypotheses regarding the thermal behavior of building envelopes. The idea of a *translation of thermal requirements into construction shapes* accurately characterizes Jean-Pierre Cordier's approach. They make it possible to explain the choice of materials and shapes.

Figure 12.4. *Les Pradettes houses (source: author's photograph)*

But while this approach was guided by thermal requirements, it also led him to design more complex systems, capable of overcoming the main problem encountered by the passive houses of the time: storage of energy in thermal form. It is for this reason that the Monchoux family home is an example:

Jean-Pierre Cordier imagined an experimental heat storage system which, at the time, was almost unknown in France.

12.3. Trapping and storing thermal energy

The problem of trapping and storing solar energy in thermal form has been a major concern since the first research on solar energy, which dates back to the end of the 19th century. The energy of solar radiation coming to the Earth is partially reflected, but a part of it is transformed into thermal energy which can be easily dissipated. In architecture, the search for storing this thermal energy has gone through various experiments on the role of materials and thermal capacity. In the first solar houses by Félix Trombe, dating from 1967 (Figure 12.5), it was a 40 centimeter-thick wall of concrete painted black which had to ensure heat conservation, this being accumulated by the greenhouse effect and transmitted by convection and conduction to the adjacent room. As hot air rises and warms up the room, heat is accumulated in the mass of concrete, which then releases it in the interior, according to a certain phase shift. In Félix Trombe's houses, the entire south facade is occupied by this heat-trapping system. This solution is architecturally very rigid.

Figure 12.5. *One of the Trombe houses at Odeillo-Font-Romeu (source: author's photograph)*

Other research projects hypothesized that it would be possible to store this heat directly into a rockbed placed under the construction raft, to take advantage of the inertia of the ground. First developed in the United States, likely at the University of Colorado in the 1950s by George Löf – a pioneering engineer in the use of solar energy – this idea was taken up by some independent home builders linked to the counter-culture of the mid-1960s, among which we can mention Paul Davis and Steve Baer, the latter being known for his constructions in the shape of geodesic domes. This technique became well-known in France in the mid-1970s and was actually tested at the end of the 1970s, namely by Pierre Lavigne, an architect and engineer who was sensitive to the use of solar energy. Jean-Pierre Cordier used a similar technique at the Monchoux family house: a rockbed was used for distributing the heat in the ground's mass, four meters deep. Here the rockbed only served as an intermediary: it was the soil's thermal capacity that was used. An important problem that this solution seemed to solve concerned inter-seasonal storage: in fact, domestic heating needs are particularly important in winter, when sunlight is weak or intermittent or, on the contrary, during periods of maximum sunlight, in summer, which is precisely when heating needs come down to zero. In theory, storage in the ground made it possible to envisage inter-seasonal storage by accumulating heat in summer and releasing it in winter. The thermal phase shift limited to a dozen hours for the wall of the Trombe house was supposed to accumulate several months with this storage system.

The system implemented at the Monchoux family home is quite complex. The greenhouse is enormous (Figure 12.6), since its function is to accumulate heat in the summer. The greenhouse is not really a livable space in summertime, although it was designed with transverse ventilation mechanisms and canopies making it possible to hide the upper components. It was not designed to be habitable during this period since its role was to trap and to transmit maximum thermal energy to the storage system. Heat is trapped in the upper area of the greenhouse and blown under the house into the rockbed buried four meters deep, via a network of conduits enabling even distribution. In the summer, this heat must be accumulated in the soil for several months. In winter, it is collected and blown into the house. Only a wood stove provides a supplementary heating function in the event of additional needs. According to the above-mentioned definition, this system could be considered passive, except for the fact that it involves a set of mechanical elements (fan, valves, etc.). In terms of solar architecture, we speak of a semi-passive system with "indirect gain", since thermal storage is not achieved directly behind the greenhouse but in a dedicated space away from the heat-trapping surface.

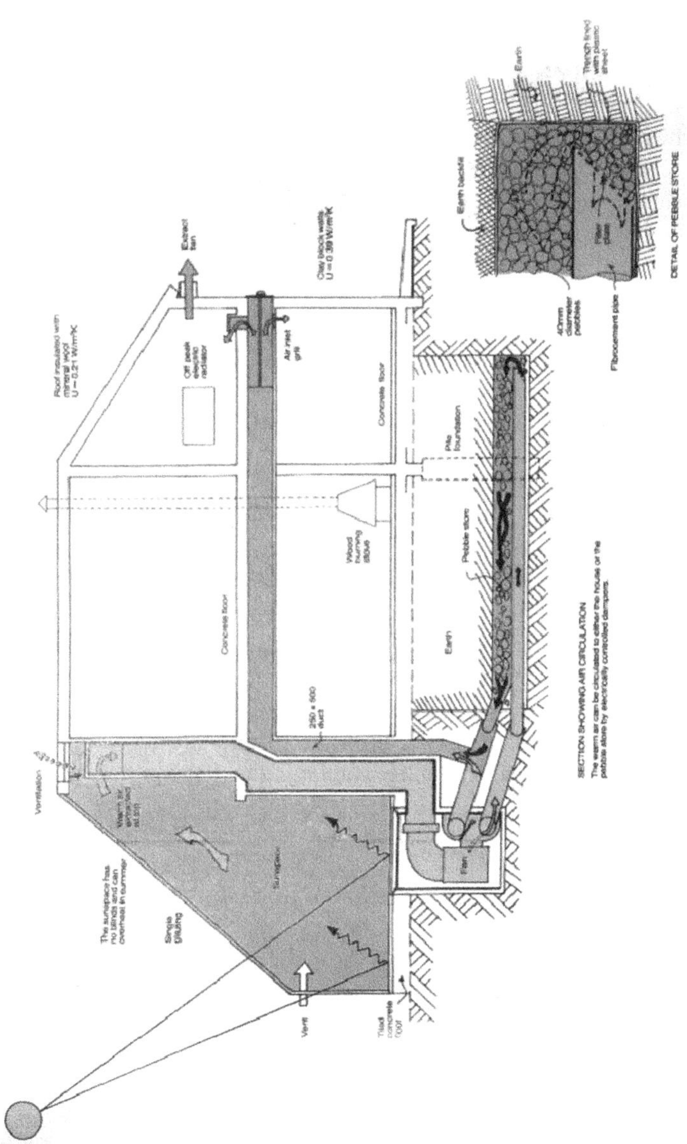

Figure 12.6. *Heat storage system in the Monchoux family home (Project Monitor, no. 7, June 1987, pp. 4–5)*

Everything we have analyzed so far concerns the theoretical functioning of the storage system; in practice, it is important to specify that this storage solution in the ground is ineffective, as it comes up against various technical problems. These technical problems are specific to this construction, whereas the rockbed system proved its worth in other works[5]. In relation to this house, the main problem is that pebbles are of irregular dimensions: the hot blown air stays blocked at the level of the air outlets and cannot heat the entire rockbed, nor does it heat the ground. In addition, since it is not isolated, there is a significant loss at the slightest difference in temperature. Considered inefficient by the dwellers and by Jean-Pierre Cordier himself, this heating system had to be abandoned, although all the mechanical elements are still functional today. Considering the house's theoretical functioning, the greenhouse maintains a purely passive role, the significant inertia of the solid concrete walls and the external insulation was doubled with a terracotta cladding. All these elements are non-negligible. Although inter-seasonal storage is inefficient, the residents, who have been the same since the house was built, are fully satisfied.

What can we learn from this inter-seasonal storage failed experience? From an experimental point of view, initiating this storage system was a particularly interesting challenge which testifies to the strength of initiative of Jean-Pierre Cordier, and more generally, for most of the first architects that embraced the bioclimatic trend. In the particular case of Jean-Pierre Cordier, he had to train himself in thermal calculations insofar as no design office was able to measure such a system. From the conceptual perspective, the idea of coupling a house with the thermal mass of the ground it rests upon seems quite original. However, the complexity of this storage system prompts us not to continue in this direction. The accumulation of mechanisms and the complexity of the storage system increase losses and risk of breakdowns. Hence, Jean-Pierre Cordier abandoned this ground storage system for a simplified approach, valuing direct gains and making the greenhouse a living space.

12.4. The greenhouse as a living space and the resident's role in bioclimatic houses

Compared to the greenhouse of the previous house, the greenhouse in the Marjolaine house, built in 1984, is very different in its layout and general shape (Figure 12.7). In the previous house, the greenhouse had a primarily technical role: it had the proper dimensions to trap the heat to be stored. Its size was that of a heat-trap directly related to the needs of the volume to be warmed up. The Marjolaine house is located in the *Pradettes* district and was built for a price per m²

5 In the constructions of Pierre Lavigne and Rober Laignelot using a similar storage system, the system works and is very efficient.

equivalent to that of low-cost housing. It reflects a change in the approach to solar heat-trapping: here, the indirect gain and the complex storage system are replaced by a direct gain. The greenhouse extends lengthwise over the entire south/south-west facade. The house is twinned with another at the level of the eastern facade (Figure 12.8), occupying one level and organized around a central living room, which can be accessed from the north.

Figure 12.7. *The greenhouse of the Marjolaine house (source: author's photograph)*

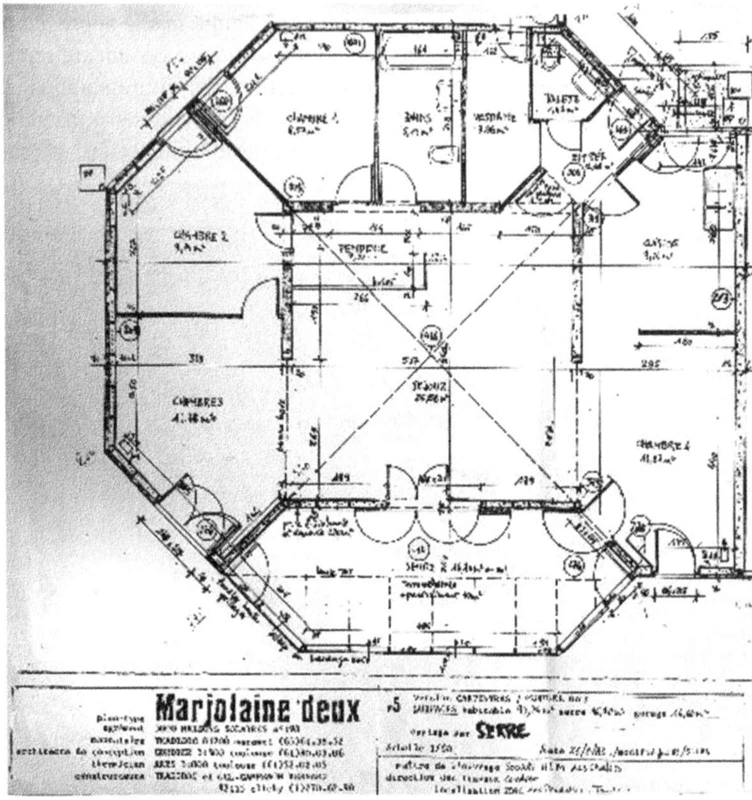

Figure 12.8. *Plan of the Marjolaine house (source: archives of Jean-Pierre Cordier)*

Needless to say, the focal point of this house is the greenhouse, which opens towards the living room. On both sides it is bordered by vegetation which creates a mask in summer, but lets light in during the winter when the plants lose their leaves (Figure 12.9). As Jean-Pierre Cordier did not plan for concealments, the owner, Mrs. Montmeyrant, modified the greenhouse by adding translucent paint in the upper area so as to attenuate radiation in the summer, when the sun is highest in the sky. For the greenhouse floor, Jean-Pierre Cordier chose interlocking paving stones. Made of concrete, the latter had some inertia which made it possible to conserve a small amount of thermal energy. But these were covered with rugs, probably because the greenhouse would have been too cold in the winter (Figure 12.7). Here we can see to what extent a choice dictated by thermal requirements can go against a certain level of comfort for the dweller, who may consciously or unconsciously prefer to do without ground inertia. Unlike concrete pavers, fabric has a very low thermal effusivity, so it feels "warm" when touched.

Figure 12.9. *The greenhouse of the Marjolaine house from the exterior (source: author's photograph)*

According to its shape, dimensions and orientation, the greenhouse fulfills a certain thermal role. Its function is to accumulate heat so as to restore it to adjacent rooms by natural convection. Therefore, a greenhouse is not a room like any other; its main function is to work as a buffer space, capable of accumulating heat and suitable for growing plants in winter. However, it requires a great deal of attention from the inhabitant to variations in the weather and to properly understand its role. Indeed, the owner says that the greenhouse is interesting on sunny days in autumn for heating the house, as long as the double door between the greenhouse and the living room is left open. The act of opening the door at the right time involves some understanding of thermal phenomena on the part of the dweller: they have an active role and control the heat input. Marie-Françoise Lavigne, wife of the architect Pierre Lavigne who was an important figure in the bioclimatic movement in France, said that the more a house was passive, the more its inhabitant had to be active: this is exactly the case in our example, since the owner must know the correct time to open the double door between the living room and the greenhouse. For instance, by opening it at the wrong time, during a winter night for example, the door would transform the house into a radiator towards the outside. As the greenhouse is single-glazed, for obvious economic reasons, the slightest difference in temperature between the inside and the outside inevitably produces a loss, not to mention linear losses[6].

Compared to the previous house where heat storage was mechanized, here the inhabitant is at the center of the house's thermal management: this supposes some basic knowledge in thermal energy and especially a sense of observation, which often make the dwellers of bioclimatic or solar houses what we could call "conscientious" or "informed" inhabitants. Sociological studies carried out on the understanding of thermal phenomena by residents have often highlighted behaviors that go against good thermal management (greenhouse heating in winter, obstruction of ventilation, etc.) (Dard 1986, p. 175). In our case, on the contrary, it seems that the inhabitants are sensitive to these issues and act accordingly. Whether it is the Monchoux family[7] or Mrs. Montmeyrant, they know the thermal behavior of their houses and how to suitably adjust the openings and concealments. While this attitude may sound normal or common sense, it is far from common[8].

12.5. Conclusion on the topicality of the bioclimatic approach

What can we learn from the creations of Jean-Pierre Cordier that can justify the title for this chapter: *Learning from Toulouse*? We have seen that experiments on storage systems have not proved any real efficiency and that simpler means, such as the greenhouse, orientation and the choice of a compact volume in particular, enabled sufficient inputs. But while the experience of inter-seasonal storage was shown to be impossible to use due to its complexity, it also enabled the greenhouse to evolve from a heat-trapping site to a living space. By providing the inhabitant with an important role in the thermal management of the house, the greenhouse at Marjolaine house becomes a greenhouse to live in. The danger of bioclimatic architecture as a form of sustainable development has been to think that complex techniques considered as economical could provide a constant level of comfort, and that this comfort could be controlled with increasing ease. However, in inhabited greenhouses it is the dweller that *modulates* the architectural elements by opening a door or closing a blind at a particular time of day, in order to maintain the interior atmosphere in its comfort zone. Modulation is essentially analog, since the

6 Linear losses refer to all the thermal losses which occur at the joints and which often result from sealing problems at the level of the frames (doors or windows).

7 It is important to point out that Ms. Monchoux is particularly sensitive to thermal phenomena, herself a physicist and personally involved in energy-related issues.

8 In order to inform the new residents about the thermal operation taking place in their home, certain architects of this period distributed small guides explaining the thermal phenomena at work, as well as instructions to be followed. This is the case with the Claux/Pesso/Raoust agency which operated during the 1980s, mainly for the "5000 Solar Houses" contest.

inhabitant acts directly on the elements which modulate comfort (doors, valves, etc.). Thermal comfort is managed through gestures and a basic knowledge of physics. As Ms. Montmeyrant says: "For me, an adequate temperature is nineteen or twenty degrees. If I feel cold, I have a blanket there and I use it while I watch TV". Comfort standards are adjusted according to variables that depend on the individuals and the habits they adopt.

What could the message of Jean-Pierre Cordier's creations be for us today? We can speak of a certain "regionalism", insofar as Jean-Pierre Cordier tried to find – through thermal and bioclimatic requirements – a form of architecture which seemed adapted to its location. The climate of Toulouse can sometimes be demanding, with rather distinct seasons and an often dry summer. The interest for us is to detect the emergence of certain recurring patterns among these achievements in Toulouse, for example, not only through the layout's shape, but also by the way in which the role of greenhouses and their function has evolved. Taking summer comfort into account to the detriment of winter comfort has also changed the shape of greenhouses, with the purpose of minimizing their harmful influence on the heat balance during hot weather. The search for regional forms undertaken by Jean-Pierre Cordier is therefore not a mere collection of picturesque elements from the Toulouse landscape, but aims to form a typology of volumes and openings relying on physical and climatic data. If we come across terracotta, in the form of "foraine" brick, Monomur or profiles, it is above all for its thermal qualities and the low maintenance it requires.

Despite the fact that some of Jean-Pierre Cordier's creations appear to be complex do-it-yourself projects, they are attempts to imagine a kind of building which is more consistent with the climate it is located in. This point characterizes most bioclimatic houses: they are houses which react to variations in climate, and to the environment, much more quickly than standard houses. A bioclimatic house is designed as *a system open to climatic inputs* which continually exchanges energy with the surrounding environment. This approach is different from the current paradigm of passive houses which maximizes insulation and minimizes thermal losses. In fact, bioclimatic constructions have a *pedagogic* dimension, insofar as they are sensitive to variations in the environment. Of course, greenhouses often overheat in the summer and are especially cold in the winter, but opening a door can mitigate this phenomenon in a matter of minutes. In addition, the presence of greenhouses encourages the inhabitant to rediscover what the architects Georges and Jeanne-Marie Alexandroff called "internal nomadism" (Alexandroff and Alexandroff 1982, p. 39), that is to say the act of occupying a room according to the seasons and its exposure. This awareness that the home is not a "well-tempered environment[9]" but an environment modulated by the inhabitant characterizes the bioclimatic approach. The inhabitant is valued in an active role, as the decisive and active center of the dwelling's thermal control. While these bioclimatic houses can be regarded as archaisms in the age of triple glazing, leakage tests and vacuum insulation, they have the merit of conveying *some understanding of climate through architectural elements* and this is perhaps the essence of their raison d'être. While it is true that nowadays we can seek to isolate ourselves from the climate, bioclimatic houses show that the climate and microclimates actually exist and can be modulated. Although the insulation paradigm is a solution in some latitudes, it is by no means the only solution to global warming. Bioclimatic houses demonstrate that the challenges of climate modulation and microclimate are topical, the essential function of architecture being to provide the means so that a climate can be inhabited.

12.6. References

Alexandroff, G. and Alexandroff, J-M. (1982). *Architectures et climats. Soleil et énergies naturelles dans l'habitat.* Berger-Levrault, Collection Architectures, Paris.

Banham, R. (2013). *L'Architecture de l'environnement bien tempéré.* Restitutions, Orléans.

Cordier, J.-P. (2013). *Conception architecturale et RT 2012. Influence des facteurs énergétiques sur la forme des bâtiments.* Le Moniteur, Paris.

Dard, P. (1986). *Quand l'énergie se domestique. Observations sur dix ans d'expériences et d'innovations thermiques dans l'habitat.* Plan Construction et Habitat, Paris.

Gaillard, C. (2020). L'approche énergétique de l'architecture vernaculaire. *Socio-Anthropologie*, 42, 67–80.

Izard, J.-L. and Guyot, A. (1979). *Archi bio.* Parenthèses, Marseille.

Jourda, F.-H. (2012). *Petit manuel de la conception durable.* Archibooks, Paris.

Mazria, E. (1979). *The Passive Solar Energy Book: A Complete Guide to Passive Solar Home, Greenhouse and Building Design.* Rodale Press Books, Pennsylvania.

Olgyay, V. (1963). *Design with Climate. Bioclimatic Approach to Architectural Regionalism.* Princeton University Press, Princeton, NJ.

Wright, D. (1978). *Natural Solar Architecture: A Passive Primer.* Van Nostrand Reinhold Company, New York.

9 In the words of Reyner Banham (1969, p. 333).

13

The !MPACT Contest: An Opportunity for Architecture Students

Building and construction are responsible for 40% of carbon emissions in developed countries. As a result, it is a matter of urgency to develop low-carbon practices in order to reduce the impact of this sector on climate change. Materials that are renewable and minimally processed like earth, straw, hemp and timber have low embodied energy and therefore constitute a response to the environmental challenges the construction sector is currently facing. The implications in regards to training tomorrow's architects to transition to a low-carbon building sector are huge.

The !MPACT student architecture contest is dedicated to promoting low-impact building materials among students and to help them develop skills and methods for working with these materials.

Today, the building sector has a very high impact on the environment. Responsible for 40% of CO_2 emissions in developed countries[1], its impact on global warming is no longer in question. The materials currently integrated into buildings are the main culprits for the sector's carbon footprint. The manufacturing process of cement, the main component of concrete, generates one ton of CO_2 per ton of cement produced; the annual production of cement itself has quadrupled in

---

Chapter written by KARIBATI to present the !mpact competition – an innovative company specializing in biobased building materials.

For a color version of all the figures in this chapter, see www.iste.co.uk/aithaddou/transition.zip.

1 *Ministère de la Transition Énergétique*, "Energie dans les bâtiments". Available at: https://www.ecologique-solidaire.gouv.fr/energie-dans-batiments [Accessed August 25, 2020].

just 30 years, now reaching over 4 billion tons[2]. Concrete is, besides water, the most consumed material in the world. This explains the need for a transition in the construction sector.

The development of bio-based materials in buildings is part of this transition. The IPCC report of August 8, 2019 discusses the benefits of harvesting wood in order "to replace emission-intensive materials, thereby reducing emissions in other sectors". Thus, the use of wood (in particular as a replacement for concrete in buildings) helps to store carbon in the long term, provided that forests are sustainably[3] managed.

But other biobased materials can be used to reduce the carbon footprint of the construction sector, thanks to their ability to store carbon in the long term. This is the case of plants, associated with the carbon-storing phenomenon of photosynthesis. Photosynthesis is characterized by the following equation:

$$6CO_2 + 6H_2O \rightarrow 6CH12O6 + 6O_2$$

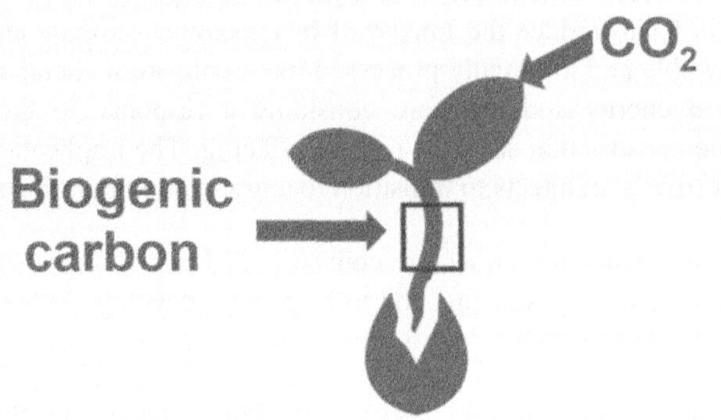

Figure 13.1. *The phenomenon of photosynthesis*

2 Vidal, J. (2019). Concrete is tipping us into climate catastrophe. It's payback time. *The Guardian* [Online]. Available at: https://www.theguardian.com/cities/2019/feb/25/concrete-is-tipping-us-into-climate-catastrophe-its-payback-time-cement-tax [Accessed August 25, 2020].

3 The IPCC defines sustainable forest management as "the stewardship and use of forests and forest lands in a way, and at a rate, that maintains their biodiversity, productivity, regeneration capacity, vitality, and their potential to fulfill now and in the future, relevant ecological, economic and social functions at local, national and global levels, and that do not cause damage to other ecosystems".

Thus, the chemical transformation which takes place during photosynthesis induces the creation of dioxygen, which will be re-emitted into the atmosphere, as well as the formation of glucose (6CH12O6) which will enable the plant to activate its growth. The carbon present in this compound is an integral part of the plant. The so-called "biogenic" carbon is the plant's constitutive carbon, resulting from the photosynthesis process of the CO_2 present in the air.

As a result of this initial removal of CO_2 from the atmosphere, plants can contribute to the reduction of "the total stock" of greenhouse gases (GHGs), and produce benefits with regard to climate change; they are said to act as a carbon well. To illustrate this concept, it is useful to consider a plot of cultivated land.

Figure 12.2 illustrates this phenomenon for 1 ha of hemp, whose annually produced biomass is integrated into a product with a 50-year lifespan. It is based on the assumption that at the end of its life, all of the CO_2 is re-emitted. Three scenarios concerning hemp cultivation are presented here:

– scenario 1: assumption of 1 ha of hemp, cultivated for 75 years, and abandoned afterwards;

– scenario 2: assumption of 1 ha of hemp, cultivated for 50 years, and abandoned afterwards;

– scenario 3: assumption of 1 ha of hemp, cultivated for 100 years.

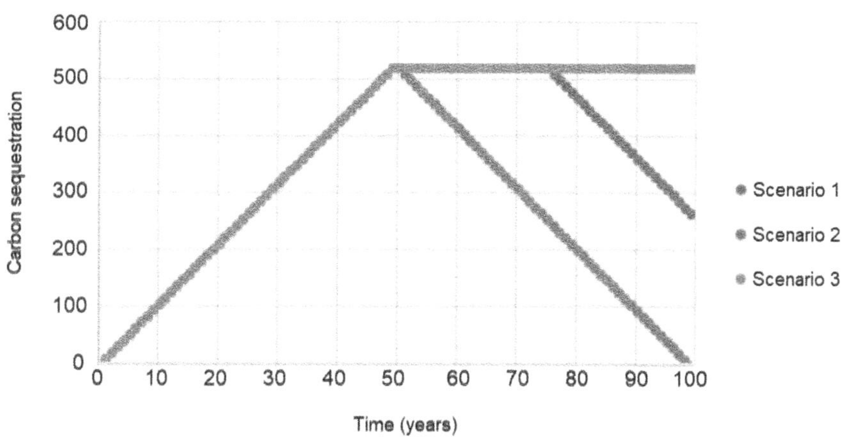

Figure 13.2. *Carbon sequestration of one hectare of hemp as a function of time, across three possible scenarios*

Therefore, biobased materials are an essential tool for reducing greenhouse gases in the construction sector. This observation raises questions about the training of architects so that they can meet the challenges of the fight against global warming.

Most architecture schools incorporate lessons on sustainable building and the city. However, during their training courses, students have little opportunity to come across project methodology which integrates all of the environmental, economic and social criteria which are required for a true transition in the construction sector. The !MPACT contest gives them this opportunity.

The !MPACT contest is a national student contest open to all students in architecture and other training fields related to building, engineering, design and art, in France and internationally. It is aimed at students of all levels of study wishing to promote sober and committed architecture.

This contest was designed to educate students and open channels of communication on ecological construction and biobased construction materials. In four years, the contest brought together more than 1,000 students and rewarded 74 students for their architectural reflection. In 2018, the contest widened its scope, promoting a "sober and committed" approach to architecture making it possible to highlight frugal materials such as bioresources, raw earth, stone, etc. The contest was also inspired by the manifesto for a happy and creative frugality, which encourages those involved in the construction industry to adopt a frugal approach, in terms of energy, materials, technicality and the territory.

The proposed project must take into account:

– energy reduction, in particular through bioclimatic design;

– controlling material consumption through the use of renewable materials and the circular economy;

– the implementation of relevant technicalities adapted to needs;

– taking into account the local scale, demonstrating an understanding of the wealth of the territory where the project is located.

The applications received as part of the contest show that students are capable of rethinking ways of rehabilitating and building, provided they are supported in their training and given the opportunity to acquire specific skills: in particular, the ability to develop a working method that takes into account a set of criteria consistent with the project's implementation. The winners, who can be viewed on the contest's website[4], offer original and relevant constructive solutions combining reuse,

4 https://www.concoursimpact.org/laureats.

geosourced materials and mostly local biobased materials. The students thus imagine prefabricated modules, new outlets for materials present locally or even new uses for unused buildings. The contest therefore enables them to develop their capacity for innovation in frugal architecture.

The contest also encourages students to seek out local pathways, in order to limit the gray energy of the materials used, as well as to develop the territory's social and economic fabric.

In addition, students are invited to present a budget to demonstrate that their project has reached level 3 of the "Biosourced Building" label[5]. This is to familiarize them with this label and enables them to understand the 2020 Environmental Statements (FDES, *Fiche de Déclaration Environnementale et Sanitaire*). *In fine*, the contest is a way of enabling them to get a step ahead of future regulations.

Finally, the past four years of the contest show that the winning projects have a particular aesthetic sensibility. Although it is difficult to attribute a reason for this, the use of materials from our living environment is perhaps a source of inspiration for the students.

5 For further information see: http://www.karibati.fr/label-batiment-biosource/ and http://www.karibati.fr/wp-content/uploads/2016/09/160817-Avis-expert-criteres-biosources-renovation.pdf.

List of Authors

Hassan AIT HADDOU
Forms, Architectures and Environments Innovation Laboratory
ENSA Montpellier
France

Edith AKIKI
ENSA Paris-Val de Seine
France

Damien ANTONI
SYVIL – Architectures du Système Ville
ENSAP Lille
France

Jérémie BUTTIN
DSAA Urban Alternatives (Higher Diploma of Applied Arts)
Vitry-sur-Seine
France

Anne COSTE
ENSAG
AECC
Grenoble Alpes University
France

André DE HERDE
Faculty of Architecture, Architectural Engineering and Urban Planning
Catholic University of Louvain
Belgium

Frédéric DELLINGER
ENSAG
AECC
Grenoble Alpes University
France

Emmanuel DOUTRIAUX
EVCAU Laboratory
ENSA Paris-Val de Seine
France

Cécile FRIES-PAIOLA
SHSA and LHAC
ENSA Nancy
and
2L2S
University of Lorraine
France

Clément GAILLARD
University of Paris 1 (UMR Géographie-Cités, CRIA Team)
France

Alain GUEZ
ENSA Nancy
France

Theodore GUUINIC
LIFAM Laboratory
ENSA Montpellier
Paul-Valéry Montpellier 3 University
France

KARIBATI
Montreuil
France

Tibo LABAT
DSAA Urban Alternatives (Higher Diploma of Applied Arts)
Vitry-sur-Seine
France

Christophe LAURENS
DSAA Urban Alternatives (Higher Diploma of Applied Arts)
Vitry-sur-Seine
France

François NOWAKOWSKI
ENSA Lyon
France

Valentin SANITAS
Conservatoire National des Arts et Métiers
UMR MAP-MAACC 3495 CNRS
Paris
France

Marjan SANSEN
LIFAM Laboratory
ENSA Montpellier
France

Antoine STECK
ENSA Nancy
France

Séverine STEENHUYSE
ENSA Marseille
France

Magalie TÉCHER
Forms, Architectures and Environments Innovation Laboratory
ENSA Montpellier
France

Dimitri TOUBANOS
EVCAU Laboratory
ENSA Paris-Val de Seine
France

Isabelle VERVISCH-FORTUNÉ
Architectural Research Laboratory (LRA)
ENSA Toulouse
France

Philippe VILLIEN
IPRAUS Laboratory
ENSA Paris-Belleville
France

Index

Other titles from

ISTE

in

Science, Society and New Technologies

2021

BARDIOT Clarisse
Performing Arts and Digital Humanities: From Traces to Data
(Traces Set – Volume 5)

BENSRHAIR Abdelaziz, BAPIN Thierry
From AI to Autonomous and Connected Vehicles: Advanced Driver-Assistance Systems (ADAS)
(Digital Science Set – Volume 2)

DOUAY Nicolas, MINJA Michael
Urban Planning for Transitions

GALINON-MÉLÉNEC Béatrice
The Trace Odyssey 1: A Journey Beyond Appearances
(Traces Set – Volume 4)

HENRY Antoine
Platform and Collective Intelligence: Digital Ecosystem of Organizations

LE LAY Stéphane, SAVIGNAC Emmanuelle, LÉNEL Pierre, FRANCES Jean
The Gamification of Society
(Research, Innovative Theories and Methods in SSH Set – Volume 2)

RADI Bouchaïb, EL HAMI Abdelkhalak
Optimizations and Programming: Linear, Non-linear, Dynamic, Stochastic and Applications with Matlab
(Digital Science Set – Volume 1)

2020

BARNOUIN Jacques
The World's Construction Mechanism: Trajectories, Imbalances and the Future of Societies
(Interdisciplinarity between Biological Sciences and Social Sciences Set – Volume 4)

ÇAĞLAR Nur, CURULLI Irene G., SIPAHIOĞLU Işıl Ruhi, MAVROMATIDIS Lazaros
Thresholds in Architectural Education (Engineering, Energy and Architecture Set – Volume 7)

DUBOIS Michel J.F.
Humans in the Making: In the Beginning was Technique
(Social Interdisciplinarity Set – Volume 4)

ETCHEVERRIA Olivier
The Restaurant, A Geographical Approach: From Invention to Gourmet Tourist Destinations
(Tourism and Mobility Systems Set – Volume 3)

GREFE GWENAËLLE, PEYRAT-GUILLARD DOMINIQUE
Shapes of Tourism Employment: HRM in the Worlds of Hotels and Air Transport (Tourism and Mobility Systems Set – Volume 4)

JEANNERET Yves
The Trace Factory
(Traces Set – Volume 3)

KATSAFADOS Petros, MAVROMATIDIS Elias, SPYROU Christos
Numerical Weather Prediction and Data Assimilation (Engineering, Energy and Architecture Set – Volume 6)

KOLOKOTSA Denia, KAMPELIS Nikos
Smart Buildings, Smart Communities and Demand Response (Engineering, Energy and Architecture Set – Volume 8)

MARTI Caroline
Cultural Mediations of Brands: Unadvertization and Quest for Authority (Communication Approaches to Commercial Mediation Set – Volume 1)

MAVROMATIDIS Lazaros E.
Climatic Heterotopias as Spaces of Inclusion: Sew Up the Urban Fabric (Research in Architectural Education Set – Volume 1)

MOURATIDOU Eleni
Re-presentation Policies of the Fashion Industry: Discourse, Apparatus and Power (Communication Approaches to Commercial Mediation Set – Volume 2)

SCHMITT Daniel, THÉBAULT Marine, BURCZYKOWSKI Ludovic
Image Beyond the Screen: Projection Mapping

VIOLIER Philippe, with the collaboration of TAUNAY Benjamin
The Tourist Places of the World
(Tourism and Mobility Systems Set – Volume 2)

2019

BRIANÇON Muriel
The Meaning of Otherness in Education: Stakes, Forms, Process, Thoughts and Transfers
(Education Set – Volume 3)

DESCHAMPS Jacqueline
Mediation: A Concept for Information and Communication Sciences
(Concepts to Conceive 21st Century Society Set – Volume 1)

DOUSSET Laurent, PARK Sejin, GUILLE-ESCURET Georges
Kinship, Ecology and History: Renewal of Conjunctures
(Interdisciplinarity between Biological Sciences and Social Sciences Set – Volume 3)

DUPONT Olivier
Power
(Concepts to Conceive 21st Century Society Set – Volume 2)

FERRARATO Coline
Prospective Philosophy of Software: A Simondonian Study

GUAAYBESS Tourya
The Media in Arab Countries: From Development Theories to Cooperation Policies

HAGÈGE Hélène
Education for Responsibility
(Education Set – Volume 4)

LARDELLIER Pascal
The Ritual Institution of Society
(Traces Set – Volume 2)

LARROCHE Valérie
The Dispositif
(Concepts to Conceive 21st Century Society Set – Volume 3)

LATERRASSE Jean
Transport and Town Planning: The City in Search of Sustainable Development

LENOIR Virgil Cristian
Ethically Structured Processes
(Innovation and Responsibility Set – Volume 4)

LOPEZ Fanny, PELLEGRINO Margot, COUTARD Olivier
Local Energy Autonomy: Spaces, Scales, Politics
(Urban Engineering Set – Volume 1)

METZGER Jean-Paul
Discourse: A Concept for Information and Communication Sciences
(Concepts to Conceive 21st Century Society Set – Volume 4)

MICHA Irini, VAIOU Dina
Alternative Takes to the City
(Engineering, Energy and Architecture Set – Volume 5)

PÉLISSIER Chrysta
Learner Support in Online Learning Environments

PIETTE Albert
Theoretical Anthropology or How to Observe a Human Being
(Research, Innovative Theories and Methods in SSH Set – Volume 1)

PIRIOU Jérôme
The Tourist Region: A Co-Construction of Tourism Stakeholders
(Tourism and Mobility Systems Set – Volume 1)

PUMAIN Denise
Geographical Modeling: Cities and Territories
(Modeling Methodologies in Social Sciences Set – Volume 2)

WALDECK Roger
Methods and Interdisciplinarity
(Modeling Methodologies in Social Sciences Set – Volume 1)

2018

BARTHES Angela, CHAMPOLLION Pierre, ALPE Yves
Evolutions of the Complex Relationship Between Education and Territories
(Education Set – Volume 1)

BÉRANGER Jérôme
The Algorithmic Code of Ethics: Ethics at the Bedside of the Digital Revolution
(Technological Prospects and Social Applications Set – Volume 2)

DUGUÉ Bernard
Time, Emergences and Communications
(Engineering, Energy and Architecture Set – Volume 4)

GEORGANTOPOULOU Christina G., GEORGANTOPOULOS George A.
Fluid Mechanics in Channel, Pipe and Aerodynamic Design Geometries 1
(Engineering, Energy and Architecture Set – Volume 2)

GEORGANTOPOULOU Christina G., GEORGANTOPOULOS George A.
Fluid Mechanics in Channel, Pipe and Aerodynamic Design Geometries 2
(Engineering, Energy and Architecture Set – Volume 3)

GUILLE-ESCURET Georges
Social Structures and Natural Systems: Is a Scientific Assemblage Workable?
(Social Interdisciplinarity Set – Volume 2)

LARINI Michel, BARTHES Angela
Quantitative and Statistical Data in Education: From Data Collection to Data Processing
(Education Set – Volume 2)

LELEU-MERVIEL Sylvie
Informational Tracking
(Traces Set – Volume 1)

SALGUES Bruno
Society 5.0: Industry of the Future, Technologies, Methods and Tools
(Technological Prospects and Social Applications Set – Volume 1)

TRESTINI Marc
Modeling of Next Generation Digital Learning Environments: Complex Systems Theory

2017

ANICHINI Giulia, CARRARO Flavia, GESLIN Philippe, GUILLE-ESCURET Georges
Technicity vs Scientificity – Complementarities and Rivalries
(Interdisciplinarity between Biological Sciences and Social Sciences Set – Volume 2)

DUGUÉ Bernard
Information and the World Stage – From Philosophy to Science, the World of Forms and Communications
(Engineering, Energy and Architecture Set – Volume 1)

GESLIN Philippe
Inside Anthropotechnology – User and Culture Centered Experience (Social Interdisciplinarity Set – Volume 1)

GORIA Stéphane
Methods and Tools for Creative Competitive Intelligence

KEMBELLEC Gérald, BROUDOUS EVELYNE
Reading and Writing Knowledge in Scientific Communities: Digital Humanities and Knowledge Construction

MAESSCHALCK Marc
Reflexive Governance for Research and Innovative Knowledge (Responsible Research and Innovation Set - Volume 6)

PARK Sejin, GUILLE-ESCURET Georges
Sociobiology vs Socioecology: Consequences of an Unraveling Debate (Interdisciplinarity between Biological Sciences and Social Sciences Set – Volume 1)

PELLÉ Sophie
Business, Innovation and Responsibility (Responsible Research and Innovation Set – Volume 7)

2016

BRONNER Gérald
Belief and Misbelief Asymmetry on the Internet

EL FALLAH SEGHROUCHNI Amal, ISHIKAWA Fuyuki, HÉRAULT Laurent, TOKUDA Hideyuki
Enablers for Smart Cities

GIANNI Robert
Responsibility and Freedom (Responsible Research and Innovation Set – Volume 2)

GRUNWALD Armin
The Hermeneutic Side of Responsible Research and Innovation (Responsible Research and Innovation Set – Volume 5)

LAGRAÑA Fernando
E-mail and Behavioral Changes: Uses and Misuses of Electronic Communications

LENOIR Virgil Cristian
Ethical Efficiency: Responsibility and Contingency (Responsible Research and Innovation Set – Volume 1)

MAESSCHALCK Marc
Reflexive Governance for Research and Innovative Knowledge (Responsible Research and Innovation Set – Volume 6)

PELLÉ Sophie, REBER Bernard
From Ethical Review to Responsible Research and Innovation (Responsible Research and Innovation Set – Volume 3)

REBER Bernard
Precautionary Principle, Pluralism and Deliberation: Sciences and Ethics (Responsible Research and Innovation Set – Volume 4)

VENTRE Daniel
Information Warfare – 2nd edition